How *to* Raise Chickens *for* Mea

How *to* Raise Chickens *for* Meat

THE BACKYARD GUIDE TO
CARING FOR, FEEDING, AND BUTCHERING YOUR BIRDS

MICHELLE MARINE

Skyhorse Publishing

Text and photographs copyright © 2020 by Michelle Marine, except page 3 © Getty Images

Skyhorse Publishing books may be purchased in bulk at special discounts for sales promotion, corporate gifts, fund-raising, or educational purposes. Special editions can also be created to specifications. For details, contact the Special Sales Department, Skyhorse Publishing, 307 West 36th Street, 11th Floor, New York, NY 10018 or info@skyhorsepublishing.com.

Skyhorse® and Skyhorse Publishing® are registered trademarks of Skyhorse Publishing, Inc.®, a Delaware corporation.

Visit our website at www.skyhorsepublishing.com.

10 9 8 7 6 5 4 3 2

Library of Congress Cataloging-in-Publication Data is available on file.

Cover design by Laura Klynstra
Cover photo credit: Michelle Marine

Print ISBN: 978-1-5107-5104-0
Ebook ISBN: 978-1-5107-5106-4

Printed in China

This book is dedicated to my grandma, Ruth Shafer, who sparked my interest in homesteading with her beautiful farm, autobiography, and pictures of her chicken flocks; my grandpa, Tom Moore, who rose above his sharecropper beginning to help integrate rural schools and told the funniest stories about his early life on the farm; my parents, Lynn and Fran Shafer, who taught me I can do anything I set my mind to; and my husband and kids, Dan, Anna, Ben, Cora, and Sara Marine, who love and support me in spite of my crazy obsession with healthy food, backyard chickens, turkeys, geese, guineas, ducks, and peacocks.

CONTENTS

How *to* Raise Chickens *for* Meat

Introduction

Thank you for picking up a copy of *How to Raise Chickens for Meat*. I'm Michelle, and I'll be your guide, providing practical tips and tricks as you raise, butcher, and cook your meat chickens. You are embarking on an emotional journey, but one that is so fulfilling. In this day and age, when many people don't even know potatoes grow in dirt, raising your own meat chickens may seem drastic and unnecessary but can be an important step in becoming more self-sufficient.

I stumbled upon raising meat chickens accidentally when a friend brought me several chicks and a few ducks in need of a good home. Neither she nor I originally knew that the chickens were meat birds. They were purchased around Easter from a farm supply store by a city-dwelling youngster who had no idea that they would grow big and stinky. Being the frugal-minded person I am, I immediately offered up my homestead when the chickens needed to be relocated. Who wouldn't want free egg-layers, after all? And so, the chicks came to live with me.

After they arrived, however, it was obvious these birds weren't egg-layers after all. They had huge feet. They were missing feathers, and it looked like they had bubbles on their chests. After a bit of research, I realized that they were actually meat birds. That's when the panic set in. How would I raise broilers? Did I want to raise these "Franken" chickens? What if they exploded on me? How would I know when they would be ready to butcher? Most importantly, how on earth was I going to butcher them when it was their time?

Not only did I need to butcher those first few chickens that I had been so happy to get for free, but around the same time, my dog mauled one of my egg-layers. She didn't manage to kill the chicken but hurt her badly enough that she needed to be put down. That's when I realized that learning how to kill and butcher chickens was not just part of raising meat birds, but it was also a necessary skill for any humane chicken flock owner.

I wish I could say the story got better from there. But the first time I butchered chickens was a total disaster. In retrospect, I suppose *total disaster* isn't really a fair description. The chickens did end up in my freezer, after all, and we ate them. However, it wasn't perhaps the most humane ending for my chickens and was more traumatic for me and my family than I'd hoped. My hope is that this book will help you have a much better outcome than my first experience!

Divided into four easy-to-navigate sections, *How to Raise Chickens for Meat* is packed with practical information. The first section, **Getting Started**, includes information on breed specifics, timing, and quantity. This section will help you analyze options and make informed decisions.

The second section, **Care & Feeding**, dives into the specifics of keeping your flock healthy. Learn how to set up a brooder, what to feed your chickens, how to safely pasture them, and how to keep your flock stress-free.

The third section, **Butchering**, prepares you for one of the more challenging parts of raising chickens for meat. It addresses some of the emotions you may feel along with the actual process of butchering and provides practical tips to make it easier. It also discusses alternative options if you don't want to process your own chickens.

The book concludes with cooking tips and delicious tried-and-true farm-to-table recipes to impress even the most doubtful family member!

If self-sufficiency and raising your own food are important to you, this book will help you pull together a complete farm-to-table experience. With beautiful photography, practical tips, check lists, and funny stories, *How to Raise Chickens for Meat* is the resource your homestead library has been missing.

Part I
Getting Started

WHY RAISE CHICKENS FOR MEAT?

I must have been about four or five years old in my earliest memories of my Grandma and Grandpa Shafer's farm in Southeast Missouri. I loved visiting my grandparents' farm. They had many outbuildings that were fun to explore: a gorgeous red barn, an empty chicken coop, a shed full of cool old things, and even a pond.

The chicken coop was always the subject of most of my speculation. You see, before my father was born (he's the youngest of six), my grandparents lived in that old chicken coop for about seven months while they built a new house. My father wasn't around to experience it, but I remember my grandma telling me stories. She also wrote about it in her autobiography, *How I Became Whatever It Is I Am.*[1]

One summer in 1945, my grandpa needed to build a new house, and they didn't have anywhere else to live, so they moved the chickens

[1] Shafer, Ruth Brown, *How I Became Whatever It Is I Am* (NAPSAC Reproductions, 1996).

out of the four-hundred-square-foot chicken house they had built themselves for $500. My grandma cleaned and fumigated, though my father assures me that her chicken coop was never allowed to get as dirty as mine, and they moved their family of seven into that coop.

My grandma had long stopped raising chickens by the time I came along, but there has always been nostalgia on my part. I loved looking at the pictures of her chicken flocks and hearing stories about how she made fried chicken from chickens she raised and butchered herself. While raising chickens for meat is a lot different now, it was partly that nostalgia that served as my motivation for wanting to try.

If you're reading this book, you're probably in one of two categories: Either you are considering raising chickens for meat, or you're like me, and you already have meat chicks at home and now you need to figure out what to do with them. Maybe you're motivated by nostalgia, or perhaps you'd like to become more self-sufficient. Maybe you have environmental or animal welfare concerns and want to avoid factory-farmed meat. Perhaps you think you may save some money by raising your own meat chickens. Or maybe you started off as a vegetable gardener and would really enjoy providing a more complete farm-to-table experience for your family. Whatever your reason for wanting to raise chickens for meat, there are several clear benefits:

- You become more self-sufficient.
- You learn skills that are mostly forgotten in the industrialized world.
- Raising chickens for meat is quicker and cheaper than raising other kinds of meat animals like cows and pigs.
- It takes relatively few supplies and space to raise chickens for meat.
- You control what your chickens eat.
- You control your chickens' living conditions.

Opting out of factory-farm meat is a main reason many people decide to raise their own chickens for meat. The National Chicken Council's Broiler Chicken Industry Key Facts 2019 online handout states that

the United States has the largest broiler chicken industry in the world and produced more than 9 billion broiler chickens in 2018.[2] About thirty federally inspected companies employ roughly 25,000 family-owned farms to raise those 9 billion chickens. These thirty companies control all aspects of commercial chicken production including raising, processing, and marketing. And they have done a fantastic marketing job, because chicken is the number one protein consumed by Americans, at the rate of 93.5 pounds per capita.

Animal welfare concerns and the environmental impact of raising large numbers of animals under one roof are very motivational for some people to raise their own chickens for meat. Raising 9 billion chickens in factory farms comes at a cost to the animals and the environment. According to the National Chicken Council, most commercial broilers are raised in huge, cage-free buildings called growout

2 "National Chicken Council's Broiler Chicken Industry Key Facts 2019," National Chicken Council. 2019 (https://www.nationalchickencouncil.org/about-the-industry/statistics/broiler-chicken-industry-key-facts/), page 1.

houses. At about 400 feet long by 40 feet wide, up to 20,000 birds are housed in a 16,000-square-foot space during their short lives. That means that each bird has only about $\frac{8}{10}$ of one square foot of space as it grows. While $\frac{8}{10}$ of one square foot doesn't sound like a lot of space, it is bigger than the Council for Agricultural Science and Technology (CAST) recommends as the minimum space allowed per chicken: only ½ square foot per bird. The chicken industry is quick to point out that their broilers are housed in very comfortable and protective environments. They're never hot or cold, they can move about freely on the floors of the growout houses, and they don't have to worry about being eating or maimed by predators. However, most of the chickens never set foot outside and they often do not have access to natural sunlight. Commercial broilers also cannot perch, nest, or forage for food like chickens naturally do.

All of these concerns led me to begin raising my own chickens for meat even though killing the chicken at the end is an emotional undertaking. For me, the benefits far outweigh the drawbacks. I think it's important to understand how we get our food and appreciate what it takes to produce the meat we eat. It's not easy to end an animal's life, but I know my chickens led a comfortable life. They spend a good portion of their life on pasture, they have plenty of room to move around, and they are allowed to perch, nest, and forage. Even though I typically raise a fast-growing Cornish Cross broiler, they still act like normal chickens. Despite the accepted industry approach to raising meat birds, I personally believe that providing a better quality of life for the birds is a more holistic and sustainable approach to food production. That said, there are also a few reasons you might not want to raise meat chickens.

WHY YOU MIGHT DECIDE NOT TO RAISE CHICKENS FOR MEAT

If you're on the fence about raising meat chickens, you should ask yourself a few questions you before committing. The first question is, do you care if it ends up costing more to raise the chickens than it would to buy them at the grocery store? Chances are, it won't be any cheaper to raise your own than it is to buy them. If your main motivation for

raising meat chickens is to save money, then this endeavor might not be for you. Whole chickens are pretty cheap at the grocery store, and it takes a lot less work to walk into the store and buy a frozen chicken than it does to raise them yourself.

Another thing you need to consider is if your family enjoys eating whole chickens. A lot of people only like boneless, skinless chicken breasts or thighs. Will you be able to get your family to eat the whole chicken? Are you able to cut a whole chicken into manageable pieces? Do you want to deal with skin, and gristle, and bones? Many cooking skills are no longer taught, and cooking an entire chicken can be an intimidating undertaking that's not for everyone.

The last question you need to ask yourself is if you will be able to butcher your chickens. If you can't butcher you own, will you be able to find someone to do it for you? It can be very emotional to end the

life of birds you have nurtured from fluffy little babies, but it will have to be done. If you don't think you can butcher them and aren't able to find someone to do it for you, then you might want to think long and hard before bringing meat birds home to raise.

RAISING MEAT CHICKENS

Now that you have decided to raise chickens for meat, you'll need to understand the basics. Where will you raise the chickens? How many will you raise? When will you raise them? What types of chickens will you raise? And, finally, what supplies will you need? You will find the answers to all those questions in the next few chapters.

Where Will You Raise the Chickens?

Most people who raise chickens for meat live in a rural setting with access to land. You don't need very much land at all to raise chickens for meat, though, and some families are able to raise chickens for meat on small lots. Before you start, you will need to check with your local zoning laws to make sure your neighborhood doesn't have ordinances against raising livestock. Many areas now allow backyard hens, but most areas only allow a certain number of chickens (five or fewer, for instance), and they usually do not allow roosters. Meat chickens are not loud, and they grow very quickly. However, they are also messy and smelly. Since many people raise straight-run meat chickens (both pullets and cockerels), there are usually roosters in the mix. Fast-growing broilers typically do not crow before it's time to butcher, but rangers and heritage breeds might reach sexual maturity and start crowing before butchering time. Some local zoning laws will allow you to raise the chickens, but not process them, so you will need to check into that, too.

While you don't need a lot of space to raise chickens, I definitely recommend having more than the $\frac{8}{10}$ of one foot of space per bird that most commercial chickens are afforded. My meat birds stay in a moveable chicken tractor and also free-range, but they are pretty lazy and do not stray far from their food. It's generally recommended to have at least one and half to two feet of space per chicken. You'll also need to know that the chickens will eat grass pretty quickly. If you want them to have access to pasture, you will need to move them every couple of days to keep them on grass. You can also keep your chickens

indoors, of course. There are no rules that dictate you must raise only pastured chickens.

Selecting the Proper Breed

The days of Grandma heading out to the coop for a chicken for dinner are mostly gone. While it is certainly still possible, it's more common for people to choose a type of chicken that has been very carefully bred to be quick-growing and feature superior meat quality. There are a few different types of chickens commonly referred to as *broilers*. Please note that while some people refer to broilers as *Franken-chickens* and swear they have been genetically modified, no chickens have actually been genetically modified. Rather, they have been carefully bred over the years to produce an efficient and quick-growing chicken that is also delicious.

Cornish Cross Broilers

Cornish Cross broilers are by far the most common type of meat chickens. Developed in the late 1940s, they are a cross between the Commercial Cornish chicken and a White Rock chicken. Cornish Cross broilers are also called Cornish X, Cornish Cross, Broiler, Cornish/ Rock, Jumbo Cornish Cross, and more. They revolutionized the meat

industry by more than halving the time it took to ready a chicken for market. Cornish Cross broilers are the most common meat bird on the market today, because they are the most efficient, fastest-growing meat bird. They are inexpensive to feed out and grow to a very uniform size at processing.

Cornish Cross broilers reach their market weight of eight to twelve pounds in a quick six to eight weeks. They are top-heavy and produce large breasts, big legs, and big thighs. They have been bred to produce more white than dark meat. Despite their reputation for being horrible birds, I have always found Cornish Cross broilers to be very sweet. They are docile and lazy, yet they still follow me around, mostly begging for food. Once they finally grow all their feathers, I think they're kind of cute. Their large legs and lack of feathers do not bother me, but a lot of people can't stand the way they look. Compared to many of the laying chicken breeds with beautiful patterns and lovely colors, it's fair to describe Cornish Cross broilers as strange looking.

It's easy to see why Cornish Cross are called Franken-chickens. Their huge yellow feet and pockets of skin without feathers can be quite alarming for the first-time meat-chicken raiser. They often have

bald spots because they grow so fast, but eventually most of their feathers do grow in. One potential drawback of the Cornish Cross is that they are not self-sustaining. This means you will have to buy them as chicks every year you plan to raise broilers; you will not be able to raise them from hatching eggs or have a broody chicken mama hatch her own eggs. Cornish Cross broilers are not broody and will not live long enough to lay or hatch their own eggs.

Cornish Cross broilers also face several health issues because of their fast growth: They are more prone to leg problems, mobility issues, and heart attacks than other types of chickens. They are lazy and sometimes can't be bothered to walk to water when they need it. Because their health issues are more common the older they get, it is sometimes possible to reduce health issues with a careful feeding schedule and butchering by eight weeks. But even with careful feeding and early butchering, I have lost Cornish Cross broilers right before processing time from apparent heart attacks. I've also had several with leg issues. They get so heavy so quickly that they just can't support their weight any longer. It is sad and frustrating to lose chickens you have raised for weeks right before their processing time, and many people want good alternatives to the broiler breed used by the commercial farms they're trying to avoid. I have heard of some Cornish Cross broilers living beyond a few months, but it's not normal for this type of chicken. People who would like a self-sustaining flock should choose a different type of chicken to raise for meat.

Rangers

If you prefer to raise meat birds that are good at foraging, don't look quite as strange as the Cornish Cross broilers, and act more like normal chickens, rangers might be a good option. These tricolored or red-feathered meat broilers are slower growing but still provide a good amount of delicious-tasting meat. Rangers are good at free-ranging and are a little bit smaller than Cornish Cross. There are quite a few different types of rangers: Freedom Rangers, Rainbow Rangers, Rudd Rangers, Red Rangers, and even Black Rangers. While they have all been bred a little bit differently by the various hatcheries, they are similar in that they are a slightly slower-growing meat-bird alternative

to Cornish Cross broilers that act more like busy egg-laying chickens instead of lazy slugs.

Unlike Cornish Cross broilers, which feed out in six to eight weeks, rangers take a little longer. That means they aren't quite as efficient at gaining weight and can be more expensive to raise. On average, rangers take anywhere from twelve to sixteen weeks to reach a mature live weight of eight to eleven pounds. Even with the longer lifespan, rangers still end up a little bit smaller than the Cornish Cross broilers. They also don't have as much white meat as the Cornish Cross, if that's important to you.

I find rangers to be a good compromise for people who object to the careful breeding of the Cornish Cross, yet still want to raise an efficient and tasty bird. Of the twenty-five rangers I recently started, twenty-two lived to processing. I butchered my rangers when they were just under sixteen weeks old, and their weights varied from three and a half to about six pounds, not quite as uniform after processing as the Cornish Cross. Also, because of their red feathers, their skin was not quite as white as the Cornish Cross, but their meat is still very delicious. One other note about my rangers: the roosters were crowing for two or three weeks before butchering.

Heritage Chickens

When you hear stories about your grandmother raising chickens, she most likely raised heritage breeds. Back in the old days, they didn't have meat chickens and egg chickens, per se. Instead, they had a flock of dual-purpose birds that provided both meat and eggs. Instead of buying chicks at the farm supply store that had been carefully bred to be quick-growing chickens, our ancestors allowed broody chickens to hatch their own eggs and raise the chicks themselves.

In her autobiography, my grandmother, Ruth Brown Shafer, describes earning spending money when she was a girl in the 1930s by raising fryers. It is quite a different process compared to how most of us raise meat chickens today.

We got "spending money" by raising our own chicks. Each of us was allowed to set three broody hens. Each hen was given fifteen carefully selected eggs that were not misshapen or had thin shells. The misshapen eggs most likely would not hatch. The thin shelled eggs would break, fouling the whole nest. The average total hatch from three settings would be 35–45 chicks. In 21 days the downy babies would pop out of the shells. Then the best natured, least excitable Biddy-Mama was chosen to mother the brood. She would be confined with the babies in a small, isolated shelter for a few days until it became home to her. Then she was allowed to roam about where she pleased. How precious they looked together!

Meanwhile the other two "settin' hens" [sic] were put in a slatted or wire coop, fed and watered for a period of not less than three days, during which time they would secede (To heck with the idea! Down with motherhood!) and begin to flirt with the nearest roving rooster. Soon they would be producing an egg a day, while the "chosen one" was knocking herself out with all that scratching, hovering, and constant clucking, trying to keep the straggling brood together.With luck we'd raise about thirty fryers each. We could expect each to bring 75 cents to $1. The feed was furnished by our parents. We did the work. The money from the fryers was our total yearly income. With the money, we bought part, or most of our clothes for the year![3]

The idyllic scene described by my grandmother motivates some people to choose heritage chicken breeds over carefully bred meat birds. According to the Livestock Conservancy, a national nonprofit dedicated to protecting endangered livestock and poultry breeds from extinction, heritage breeds are the traditional breeds that our ancestors raised. Our ancestors chose specific breeds because they were "hearty, long-lived, and reproductively vital birds that provided an important source of protein to the growing population of the country until the mid-20th century."[4] With today's commercialized agricultural sector,

---

3   Shafer, Ruth Brown, *How I Became Whatever It Is I Am* (NAPSAC Reproductions, 1996).

4   The Livestock Conservancy (https://livestockconservancy.org/index.php/heritage).

and fewer people raising their own farm animals, many of the different types of farm animals are in danger of going extinct. The commercial meat industry primarily raises Cornish Cross, and many other chicken breeds are facing struggling populations.

Heritage breeds are often more self-sufficient, because they are able to reproduce naturally, forage, and become good mothers. Some breeds also have natural resistance to diseases and parasites, and some breeds are better suited for hot or cold weather than other types of chickens. The preservation of heritage breeds is a noble and necessary undertaking; however, raising heritage chicken breeds for meat does have some drawbacks. Heritage breeds are slower-growing birds, not reaching market weight until they are at least sixteen to twenty weeks old. They produce a leaner meat than commercial market birds. Additionally, they have a less-uniform weight gain, with some birds growing quite large and other birds remaining relatively small. Because they take longer to raise, and gain weight less uniformly, they can be more expensive to raise. Heritage chicken breeds can also cost more money to purchase as chicks compared to the cheaper Cornish Cross broilers.

Common heritage breeds often used as meat birds include White Rocks, Barred Rocks, New Hampshire Reds, Delawares, Rhode Island Reds, Black Australorp, Jersey Giant, the Naked Neck Turkens, and more. My grandmother raised Rhode Island Reds, which motivated me to include the breed in my first chicken flock, as well; however, I never did butcher any to eat. I only kept them as egg-producing chickens.

The benefits of raising heritage breeds include a more natural, self-sustaining chicken flock. Letting broody hens raise chicks like my grandma did can mean less work for you. Mama Chicken gets to be the brooder, babysitter, and heater, so your first three weeks of chicken care might be easier. However, it's likely that you will have to tolerate more loss, as Mama Chicken might not be as careful or consistent as a human if you decide to let a broody hen raise a flock. Raising a dual-purpose flock of heritage breeds could be more like the scene my grandma described from her youth. Well-tempered broody mamas hatch eggs and tend the chicks. You might keep the pullets for egg-laying and cull the cockerels to eat just like they used to do in the old days. Many people find raising heritage breeds for meat is a more natural undertaking than ordering chicks through the mail.

To summarize the breeds, here's a quick chart on how they all compare.

	Cornish Cross	Red Ranger	Heritage Breeds
Time to Maturity	6–8 weeks	12–16 weeks	16–20 weeks
Mortality	High	Medium	Low
Foraging	None	Good	Best
Cost	Low	Medium	High
Finished Meat	Large breasts. Chicken shape most comparable to grocery store.	A little more dark meat, more chicken flavors.	Tougher meat due to longer foraging. Best for stews and soups.

Determining How Many Chickens to Raise

It sounds like a pretty simple question, but how do you figure out how many chickens you want to raise? To answer the question, you'll need to consider how many chickens your family will eat, how much storage space you have, and how many chicks you think you can care for, and butcher, at once.

My family of six eats on average one whole chicken per week. That means we need to raise about fifty chickens per year. Many families will raise two batches of twenty-five meat chickens twice a year to end up with about one chicken per week. It's easier to raise and butcher twenty-five chickens at a time than it is to raise fifty at once. Your meat will also be fresher if you eat all twenty-five chickens from the first batch before raising the second batch. Plus, freezer

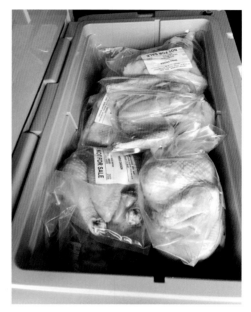

space can play a key role in determining how many chickens to raise. A whole farm-raised chicken can take up quite a lot of space. You'll definitely want to assess freezer space before buying.

If you end up with more chickens than freezer space, there are a couple things you might be able to do. The first is to cook your chickens and freeze the cooked meat. Cooked frozen chicken takes up considerably less space in your freezer than a whole chicken, especially if you pack it flat in freezer bags or vacuum seal it. The other thing you may be able to do is beg, borrow, and steal freezer space from relatives. I've stored chickens in my mother's freezer before, and I know my sister-in-law stored chickens at her parents' house. The trick is getting them to agree to this, and then remembering that you have chickens there.

Timing—When to Raise Chickens for Meat

The most common time to raise broilers is in the late winter or early spring. Since excessive heat is especially hard on chickens, most people try to beat the heat when deciding on timing. If you're buying chicks from the farm supply store, you're limited to when they sell chicks. Many stores sell chicks from March through May. Start watching your store for signage in late December or early January so you don't miss any ordering deadlines. Specific breeds sell out fast sometimes, so if you want a specific breed, it's a good idea to order them early in the season, or availability could be a problem.

The other timing consideration is processing. If you choose to outsource butchering, you will need to double-check available processing dates. Sometimes lockers only process chickens for a certain time period, usually from late spring to midsummer. Our local processing facility only butchers chickens until mid-July and then they're finished for the year. It might be a good idea to talk to your locker about their processing schedule before ordering your chicks. At the very least, call them to schedule as soon as you bring your chicks home.

If you choose to process at home, take a careful look at your schedule. If you know you'll be gone on vacation or have busy events scheduled for a specific time in the summer, make sure your chickens will not need to be processed during that time.

Where to Buy Your Chickens

After deciding which type of chickens to raise, it's time to figure out where to get them. If you choose a faster-growing Cornish Cross or one of the various rangers, you have two choices: buy chicks at a farm supply store or order them online. Whichever method you choose, make sure that you source them from a National Poultry Improvement Plan (NPIP) accredited program.

The NPIP is a voluntary certification program that was established in the 1930s to safeguard poultry from pullorum disease, which is caused by *Salmonella enterica Pullorum*. Pullorum disease was a large problem in poultry flocks in the 1930s and, according to the NPIP, was a severe threat to baby chicks, killing upward of 80 percent of

flocks.[5] A NPIP-certified hatchery regularly tests the blood of their breeding flock to make sure they do not transmit diseases like salmonella, H5/H7 avian influenza, fowl typhoid, and other diseases. Any reputable hatchery from which you wish to buy chicks should display their NPIP number on their website.

Buying Chickens at Farm Supply Stores

Another option is to buy chicks from a farm supply store. Many farm supply stores stock live chicks in the late winter or early spring. Often called a store's Chick Days, you'll start seeing signage in your store before the chicks arrive. Many stores also allow you to preorder chicks. The choice between preordering chicks through a store's sales flyer or buying them off the floor is your call, but here are considerations for each type of purchase.

If you preorder chicks, the store will give you a date and then call you when they arrive. You get to the store as quickly as you can and take your chicks home with you right away. Your chicks are most likely stored in their own box and not put out on the store floor with the rest of the chicks. When you preorder, you can also specify a straight run, all female, or all male if you have a preference but you may have to meet a minimum order quantity. Preordering might be a good option if you want to feed your chicks a specific food. If you buy chicks off the store floor, they will already be eating and drinking the store's choice of food. If you want your chicks to eat only organic, for instance, you will want to preorder and pick them up quickly so you can get them started on the right food.

You don't have to preorder chicks from most farm supply stores, though. You can also pick them up from their main store orders. If you don't care what the chicks eat for the first several days, you can also pick them up a day or two after they're delivered to the store. This could be a benefit, as it's common to lose a chick or two in the first couple of days. If you buy chicks that are already a couple days old, the less hardy birds may have already been culled. Buying birds off the

---

5   National Poultry Improvement Plan. (http://www.poultryimprovement.org/).

shelf might result in buying fewer chicks and allows you to pick them up on your own schedule.

Ordering Chicks Online

Another option for buying meat chicks is to order them online from a NPIP-certified egg hatchery. When you order chicks online, the hatchery hatches the eggs and sends the chicks through the mail. Since they're coming through the mail, you might have to order in larger numbers (sometimes at least twenty-five chicks at once) for them to keep each other warm. The US Postal Service requires that the chicks arrive at their destination within seventy-two hours of being mailed. They are sent in reinforced boxes with venting holes.

You might be wondering how chicks survive and what they eat while they're being mailed! Well, chicks actually survive in the egg by eating the nutrient-rich yolk. Eating every bit of this yolk before they hatch gives them enough nutrients to survive for a few days without any other food. Murray McMurray, a chick hatchery from Iowa, has been mailing chicks with very little mortality since 1919! As long as they're mailed as soon as they hatch, chicks can survive without any additional food or water for seventy-two hours.

Online hatcheries sell a wide variety of chicken breeds including Cornish Cross, rangers, and heritage breeds. They also often sell what they call a "fry plan rooster special." Since a lot of people prefer only female chicks because they become egg-layers, hatcheries have to do something with the roosters. Selling heritage breed roosters as meat birds is a great way to repurpose what would otherwise be discards. They are also very economical to buy. The cost per chick goes down the more chicks you buy at a time. To save money on chicks, consider ordering in a group with other people who want to raise chickens for meat. For most hatcheries, you will pay the smallest amount per chick if you can order in quantities of 100-plus chicks.

Should You Vaccinate Your Chicks?

Marek's disease is highly contagious viral disease that affects chickens in all areas of the country. It is especially problematic in industrial settings where chickens have tight living quarters. Marek's disease is caused by a chicken herpes virus and there is no cure. Once Marek's disease is introduced to a flock, it is very difficult to remove. According to the Pennsylvania State Extension Office, Marek's disease causes "inflammation and tumors in the nerves, spinal column and brain. In this form, birds will become paralyzed in the legs, or wings or may develop head tremors."[6]

Most hatcheries offer a Marek's disease vaccination for chicks and recommend that egg-laying flocks receive the vaccine. Given the short lifespan of meat chickens, vaccinating is not generally recommended.

6   Wallner-Pendelton, Eva, and Gregory P Martin, "Marek's Disease in Chickens," Penn State Extension. August 27, 2018 (https://extension.psu.edu/mareks-disease-in-chickens).

However, if you have had a flock with Marek's disease before, you will want to make sure the new chicks do not come in any contact whatsoever with the sick flock or their living areas, or use any of their feed or water containers. Since the disease lives for months and is transferred very easily from chicken to chicken, you will also want to make sure your broilers do not live in a contaminated space.

Picking up Mail-Order Chicks

Picking up chickens from the post office is always a lot of fun. We weren't sure what to expect the first time we ordered chickens through the mail, but we really enjoyed the process. We were excited to get the shipping confirmation that our chicks were finally on the way so we could get everything ready for their arrival. It's a good idea to call the post office the week your chicks are supposed to arrive to let them you know you are expecting a live delivery. Make sure they have your correct phone number for notification, as they'll call you early on arrival morning.

The night before our first chicks were to arrive, I made sure to turn on my cell phone ringer, because it's usually on silent. It's a good thing I did, because the post office called before 7 a.m. to let us know they had a cheeping box of chicks waiting for us! The drive to the post office to pick up a box of chicks is such a fun one!

We were always thrilled when the chicks arrived on Wednesdays, because my kids have a late start at school every Wednesday. Of course, they always want to go along to the post office to help pick up the birds. Once arriving at the post office, we have to ring the bell on the back door, because they aren't open when we get there. Hatcheries recommend picking up chicks as quickly as you can because they will be hungry, thirsty, and ready to get out of their box. I'm sure your post office is different than ours, but picking up chicks can be a fun adventure you wouldn't have in the freezer section of the supermarket.

Raising Meat Birds and Egg-Layers Together

While you can raise meat chickens and egg-layers together, there are a few problems you will need to address. Since meat chickens are bred to eat, that's what they'll do most of the day. They'll even lay down to eat the food if they can, which means they're hogging all the food. If you raise meat birds and egg-layers together, you'll need to make sure the egg-layers also have access to food and that the meat chickens are not preventing them from eating.

Another thing to consider is that meat chickens and egg-layers might not need the same food. Meat chickens need a higher protein content than egg-layers. Ensuring the two types of chickens eat the proper feed can be challenging if you're raising them together. Putting the two types of birds together in the same brooder is also not a great idea, mainly because the meat chicks will grow at a much faster rate than the egg-layers. Again, the meat chicks will hog all the food and might make it difficult for the young egg-layers to eat. A last consideration about raising meat chickens and egg layers together is that meat chickens are messy, messy creatures. They poop almost as much as they eat. This makes for very dirty coops. If you don't want to clean out your egg-laying coop every couple of days, it might be a good idea to separate the two types of chickens.

Part II
Care & Feeding

Oh, happy day! You finally decided which type of meat chickens to raise. You ordered the perfect chicks and have a confirmed delivery date. Now it's time to prepare for their arrival and learn how to take care of the cute little balls of fluff! Before your chicks arrive, you'll want to have everything ready and waiting.

SETTING UP THE BROODER & HEATING REQUIREMENTS

The first thing to do is gather supplies and set up the brooder. You don't need a lot of supplies, but I recommend setting up everything several days before your chicks are supposed to come, because sometimes dates slip. I've had chicks arrive two days before they were supposed to, and I've had chicks arrive two weeks late. Just to be safe, it's better to be ready ahead of time than scurrying around at the last minute.

Where to Set up the Brooder

As you choose a location for the brooder, there are a few considerations to keep in mind. First, baby meat chicks need a very warm,

draft-free location for the first few weeks. Depending on when your chicks arrive, it might still be quite cold outside. They'll also need a lot of attention the first few days, so putting them in an accessible location is a good idea. You'll have to make sure they don't have pasty butt (more on this later), and you'll need to ensure they're eating and drinking enough. I always like having my chicks close by for at least the first week so I can check on them easier.

Over my husband's objections, I put my chicks in the basement of my house for the first couple of weeks. It's warm, draft-free, and close by for quick checks. It's also easy to fill up food and water. While ease of care is important, the third consideration is that meat chicks start to smell pretty quickly. No matter how cute and fluffy they are in the beginning, or how often you change bedding, they still poop, and poop smells, and meat birds poop more than egg-layers. Despite my husband's grumblings, the chicks stay in our house where it's easiest for me to care for them for the first several days. You could also set up a brooder in an outbuilding or garage as long as you can keep it warm and cozy. As long as you provide a warm, draft-free space that is easy to keep an eye on, you'll be in good shape!

Gathering Supplies
- Brooder box, round or oval if possible
- Heat source
- Bedding for baby chicks
- Food and water containers
- Electrolytes
- High-protein feed, 22% to 24% protein for the first three weeks
- Thermometer and mesh wire for a brooder box lid

Brooder Box

Before your chicks come home, you'll need to set up a brooder box. You can use wooden boxes, plastic totes, or some sort of large water trough. I generally start my chicks in a large plastic tote or a high plastic watering trough. High walls are a plus, because the chicks start to jump as they grow. Even lazy meat chicks roost on the sides of the brooder box—which can get messy—or jump out, which can be harmful to the chicks. If possible, it's good idea to have a round or oval container for the chicks. Meat chickens are not the smartest animals, and they have a tendency to trample each other if they get scared. Having rounded corners helps them avoid trampling.

Heat Source

Along with the brooder box, you'll need a heat source. Even in warmer climates, baby chicks need a consistently warm environment until they've lost their chick down and grown their feathers (usually around six weeks of age). There are several different types of heat sources you can provide for your chicks. The most common is the heat lamp and heat bulb. This is a metal lamp reflector with a guard and clamps you can use to hang heat-emitting light bulbs. Heat lamps and bulbs provide a lot of heat, but they also have some drawbacks you need to be aware of if you choose them for keeping your chicks warm.

First, the heat lamps can get very hot. I have burned my hand slightly by touching them when I should have known better. If your children help with the chicks, you will need to remind them to be very careful around the lamps and bulbs. The high heat also makes them a potential fire hazard. Many people have accidentally burned down their barns because of fires caused by heat lamps in the chicken coop. Heat lamps and infrared bulbs aren't very energy efficient and can increase the cost of raising chickens for meat. Finally, the wrong light bulb can stress your chicks by keeping them awake. That said, even with all these drawbacks, I use heat lamps and infrared bulbs with my own chicks. They are relatively inexpensive to buy, and other than the one slight burn, I have never had a problem with heat lamps over my brooders.

If you decide to use a heat lamp and light bulb, there are a few types of bulbs you can buy. Many people highly recommend using ceramic reptile bulbs for their chicken brooders. Because ceramic bulbs only put out heat and not light, your chickens get to experience the normal daylight and darkness that they need for proper development.

If you don't want to use a ceramic bulb, you can also use a 250-watt red light bulb. These bulbs are a little less expensive than the ceramic bulbs, and their red-light output may also reduce stress on the chicks. The red light may also discourage chicks from pecking at each other. The one type of light bulb you should not use is a white light bulb. The white light deprives your chickens of the normal day/night period they need for proper development. It also can encourage them to peck each other.

You'll also need some way to hang the lamps. I've used all kinds of strange setups for hanging lights, including an old camera tripod, and clipping the lights to shelves near the brooder boxes. Whichever way you choose to hang the bulbs, make sure it's very secure and not in any danger of falling down on the chicks. You certainly don't want your heat lamp to cause a fire.

If you prefer to forgo the heat lamp and bulb and use a different type of heater, you do have options. Several low-wattage, energy-efficient, safer heaters are available to help keep chicks warm. The major downfall with these options is that they are more expensive than the heat lamp and infrared light bulbs. Adjustable brooder heating plates, hanging coop and brooder heaters, and wall-mounted low-wattage heaters are all options you may want to explore as safer alternatives for keeping your chicks warm. If the safer products give you a better peace of mind, then that totally justifies the higher purchase price. And perhaps you will even end up saving money over the long run by consuming less energy.

Bedding for Baby Chicks

Now that you've decided how to keep your chicks warm, you'll need bedding for the brooder. The best type of bedding is nonslip and absorbent, which helps keep chicks safe, especially in the beginning while their spindly little legs are growing. We've already established that meat chickens really poop a lot. If you've never raised meat chickens before, you might be surprised by just how much comes out of their bodies. It's very important that you keep the brooder clean and dry, because disease and bacteria love wet and dirty places. Choosing the proper bedding will also help you keep your brooder clean.

So, what's the best type of bedding? There are a few good options: pine shavings, sand, and straw or hay. There's also one option that is often recommended but actually not good: newspaper. Newspaper is too slippery for little chicks. While it's free and easy to come by, it really doesn't make good bedding for your chicks. Pine shavings are a great type of bedding because they are absorbent, plentiful, and cheap. However, sometimes chicks eat the pine shavings, which can cause problems. I have not had this problem with my chicks, but if you'd like to be cautious, first layer pine shavings on the bottom inch or two of the brooder, and then cover the shavings with nonslip kitchen shelf paper or paper towels. Covering the shavings can keep the chicks from eating them, but also gives the chicks the traction they need so they don't slip and injure their little legs.

My bedding of choice is simply pine shavings. If you use pine shavings, make sure to buy untreated pine shavings made for pets. They are readily available at farm supply stores and relatively inexpensive. Plus, you probably won't need more than one bag. Another benefit of pine shavings is that they compost very well and can be used in the garden afterward. Do not buy cedar shavings, as the oil in the cedar is harmful to chicks. A lot of people really like construction-grade sand for the brooder, as it absorbs moisture well and desiccates poop instead of retaining it. It can be also be good for long-term chicken foot and nail health, which may be more of a concern for egg-layers. I don't like that it doesn't compost, so I don't personally use sand in brooder boxes. Straw is also a bedding option that I use in the chicken coop for full-grown egg-laying chickens. I find it unwieldy and hard to manage in the smaller area of the brooder box, though, so pine shavings remain my preferred bedding.

Food and Watering Containers

Your chickens will also need food and water containers. Consistent access to food is essential to growth. There are lots of different types

of feeders you can buy, but I've found the long ones with multiple openings work well in the beginning. If you choose the small round feeders, you may need more than one feeder to ensure that all the chicks have good access to food.

Water containers are as important as food containers. Again, you'll want to make sure that all the chicks can access the water. If you have many chicks, you may need more than one waterer. Waterers come in many sizes, but my favorite for meat chicks are the medium-sized waterers that hold half a gallon. Small quart containers also work but need to be filled up too often. Since the half-gallon waterers last longer, I don't have to worry about the chicks running out of water in between checking in on them.

One more consideration for the food and water containers is that the chicks move around a lot and make a mess with the food and water. It helps the dishes and bedding stay cleaner if you elevate the containers a little bit off the bedding. It's fun to watch the little meat chickens play around in the brooder, but it is a little annoying to find the food wasted in the bedding or the bedding ruined by water due to their antics. An easy way to elevate the food and water dishes are to put them on bricks in the brooder. Alternatively, you can put them on a small baking sheet or a piece of plywood. If you decide to elevate the dishes, however, make sure that you do not elevate them too high. The chickens must be able to reach both their food and their water.

Electrolytes

Another important supply you will need is a source of electrolytes. As with human electrolytes, chicken electrolytes help chicks rehydrate and rebalance when they're ill or stressed by replacing sodium, potassium, and bicarbonate that they need. Since shipping is stressful on chicks, you will need electrolytes the first day they arrive. You can either buy premade electrolytes or make your own. A common store-bought electrolyte powder comes in individual packets and is premeasured and ready to add to a gallon of water. Ingredients include potassium chloride, sodium citrate, dextrose, sodium bicarbonate, sodium chloride vitamins A, E, D, C, and seven B vitamins. Sav-A-Chick is a handy way to quickly and easily make a gallon of electrolyte solution for your chicks. It does have artificial colors, though, so if that's something you would like to avoid, you can make your own powder to keep on hand. Lots of electrolyte recipes are available online.

High-Protein Feed

Not only will your chicks be thirsty when they arrive, but they will also be hungry. Make sure you have a high-protein (20% to 22%) chick starter feed on hand for the first three to five weeks. You can expect each chick to eat roughly fifteen pounds of food over an eight-week lifespan, so if you have twenty-five chicks, you'll need around 375 pounds of food. I don't recommend buying all your food at once, as protein requirements change as the chicks get older, and it can be challenging to store that much food. Just beware that you will need quite a lot of food over your chicken's lifespan.

What to Feed Your Broilers

You will also need to decide what type of food to feed your broilers. Some people like conventional, nonorganic chicken feed. This is the cheapest type of chicken feed you can find. It's readily available at farm supply stores in twenty-five-pound bags, forty-pound bags, and sometimes fifty-pound bags. If you live near a feed mill that mixes rations for farmers, you can also buy feed directly from them in fifty-pound bags. Feed mill mixes are often my cheapest source of feed.

Other types of chicken feed include non-genetically modified (GMO) chicken feed; organic (non-GMO) chicken feed; soy-, corn-, and canola-free organic feed. If you chose one of these types of feed, you will probably have to research availability, as it can be difficult to find. My local feed store special orders organic chicken feed for me for a better price than I can get elsewhere and then I pick it up at their mill. If you have a local feed mill, it may be a good option for finding the type of chicken feed you want.

The last question you will need to ask about chicken feed is whether you want medicated or nonmedicated chicken starter. Medicated chicken feed contains an FDA-approved drug called Amprolium used to protect chickens against coccidiosis. It's important to note that Amprolium is a thiamine blocker, which means it inhibits the uptake of vitamin B1; it is not an antibiotic. There is no withdrawal period from the Amprolium, and it's considered safe to feed broilers until processing. If your chicks are vaccinated against coccidiosis on delivery, do not feed them medicated starter, as it will interfere with the vaccine and render them both ineffective.

What Is Coccidiosis?
Coccidiosis is an intestinal parasite that is spread from the dirt or infected chickens and is the leading cause of death in young chickens. People who are raising more than fifty chicks at time, have had a confirmed case of coccidiosis before, or are raising consecutive batches of meat chickens are recommended to feed their chickens medicated chicken feed. I prefer nonmedicated feed for my small-quantity, once-a-year approach to raising broilers, but you may want to make a different choice.

Money-Saving Tip on Chicken Feed:
Just as you can save money on chicks if you go in on a group order, you can also save money on food by ordering in bulk. Sometimes the only way to get specialty feeds is through a wholesale order of one ton or more. Since many backyard chicken raisers aren't raising that many chicks, ordering in bulk with friends might be a good option.

Optional: Thermometer and Mesh Wire for a Lid
The last couple of supplies that you may want to have on hand before your chicks arrive are a thermometer and a mesh wire lid for the brooder box. The first time I raised chicks, I had a thermometer in

the brooder box, which helped me keep the brooder at a consistent temperature needed by the chicks. However, I've since learned to rely on my chicks to tell me if they're hot or cold. It's pretty easy to tell by their actions. If chicks are huddled together under the heat lamp, they're too cold. If they're spread out trying to get away from the heat lamp, they're too hot. If they are scattered uniformly through the brooder box, they are just right. You can easily change the temperature in the brooder box by raising or lowering the heat lamp. A mesh wire lid for your brooder box might also be a good idea. If your brooder box has short sides and it looks like the chicks will be able to hop out, a mesh wire lid can help keep them inside. It can also keep other animals out of the box if that is a concern.

Heating Requirements

Baby chicks need a lot of heat and can die if they are cold for too long. Having two heat lamps in the beginning is always advised so you have a spare if one bulb stops working. That said, too much heat is also dangerous. Not only will the warmth of the room contribute to keeping the chicks warm, but the chicks themselves also give off heat. The more

chicks you have, the more heat they'll give off. It's a good idea to put your heat lamps off to a side in the brooder, leaving a cooler place to go if they start to get too hot. In the first few days, watch your chicks closely to make sure they're warm enough, but also to make sure they don't get too hot.

Here are the heating requirements for baby chicks. The temperature should be reduced by one or two degrees every day by raising up the heat lamp. As the weather gets warmer and the chicks grow bigger, you will need less and less heat. It's highly unlikely that you will need artificial heat sources for more than three or four weeks.

Week 1 = 90°F–95°F
Week 2 = 85°F–90°F
Week 3 = 80°F–85°F
Week 4 = 75°F–80°F
Week 5 & 6 = 70°F–75°F
Weeks 7 & 8 = 65°F–70°F

WHAT TO DO THE DAY YOU GET YOUR CHICKS

The day you get your chicks is a very exciting one! Hopefully your brooder is set up and waiting, the food containers are full, and the water containers are ready with electrolytes. When you get your chicks home, it's important to gently dip each chicken's beak in the water. Dipping the beaks in water serves two purposes. First, it encourages the chicks to take their first drink. If they came in the mail, they're likely thirsty. If you picked them up from the store, they might be a little stressed from their travels, as well. Second, dipping their beaks in water helps them learn where the water is so they can get a drink when they're thirsty.

One by one, dip each chick's beak in the water and gently place it in the brooder. Then sit back and watch them for a little while. Baby chicks will run about and play. They might eat some food, and they will sleep. Don't be alarmed if it looks like your chicks are dead when they're sleeping, as sometimes they just seem to fall over and sleep where they land. Chances are they are just fine! Observe them for a little while, though, to make sure they seem warm enough and to make sure they can't trample each other in the corners of your brooder. You'll also want to make sure their brooder area is not too big so they will stay in the warm spot and not freeze to death in a corner. If you notice them wandering off, add a partition to keep them closer to the heated area. The old saying about chickens being birdbrains is pretty accurate. Meat chickens are not the smartest birds.

TAKING CARE OF MEAT CHICKS THE FIRST FEW WEEKS

The most important tasks the first few days of keeping meat birds is to make sure they have clean, cool water, access to food, and that they are warm and dry. Meat chicks should always have access to clean, cool water; they should never run out. They should also have access to food 24/7 for the first few days, so keep filling up their food, as well. You will need to be vigilant about keeping their food and water containers clean. Pop in to check on your chicks several times a day the first few days you have them to make sure they're getting along okay. If it looks like they are too hot or too cold, adjust their lights until they act just right.

Be Vigilant about Pasty Butt

Also, over the first few days, you will want to be on the lookout for pasty butt. Caused primarily by chilling or overheating, pasty butt is more common in chicks that come through the mail than chicks that are hatched and cared for by a mama hen. Pasty butt is a condition where soft poop sticks to a chick's behind and clogs the vent as it hardens. Most chicks are able to poop without issues and keep their behinds clean, but every once in a while, you will spot a chicken that has a dirty bottom. This crusty mess needs to be dealt with quickly, as it can clog your poor chick and end up killing it.

You'll probably only have to be on the lookout for pasty butt for the first few days after you get them. It seems to clear up on its own

after a couple of days. If you find chicks with poopy bottoms, make sure to remove the hardened poop. If you catch it quickly, you might be able to simply wipe the poop off. But if it's already hardened over the vent, you will need to wet the chick's behind and clean it off.

Start by running a little warm water over the chick's bottom, then carefully and gently wipe the dried poop off using an old kitchen towel or a paper towel. If it's really hard, you might have to repeat the process a few times to thoroughly clean the chick. Make sure to use warm water, not cold water, to avoid chilling. The chick will probably not enjoy being handled, so work gently and efficiently, holding her wings steady so she can't flap around. Dry the bottom the best you can and put the chick back in the brooder. If you accidentally injure the chick's skin when you remove the poop, put a little salve on the chicken to protect the area, and watch the chick to make sure the other chicks don't peck at her. To help prevent pasty butt, make sure the chicks are not overheated or chilled. It's also helpful for your chicks to be drinking well, so make sure they have clean, cool water and encourage them to drink. If pasty butt is a persistent problem, you might consider switching feed brands altogether to see if that helps. Practice good animal husbandry.

In addition to keeping the brooder warm, you'll also want to keep it clean and dry. You might be surprised by how much water and food those little chicks are capable of tracking all over the place. To keep the brooder in tip-top shape, clean out the bedding as often as necessary. To clean the bedding, carefully move the baby chicks to one side of the brooder box, scoop out the old bedding, and put fresh bedding down. Elevating the food and water dishes can help keep the brooder box cleaner. If you don't want to remove the bedding, you may be able to put fresh bedding on top of the old bedding and layer it.

TAKING CARE OF CHICKENS FROM WEEK FOUR

It's a good idea to get your broilers out on pasture as soon as possible, generally when they are between two to four weeks old or as soon as they have grown in their feathers and the weather is warm enough for them to be without a heat lamp. If you grow a garden, you know that you can't just plant seedlings outside without hardening them off first, and chicks are the same. You will need to acclimate them to the cooler temperatures and wind by opening windows and reducing temperatures in the brooder before putting them outside. If you take your chicks straight outside from the warm brooder, they are likely to be stressed, and stressed chickens are not happy chickens. As you decide when to put your chicks outside, it's important to watch the weather. Choose a calm day to take them out. Putting them on grass right before a big rain- or windstorm can stress out the chickens. Chickens can die from exposure to the elements before they are ready, so be careful when taking them outside. Gradually increase exposure over a few days until your chicks are comfortable in their new environment.

Housing Chickens on Pasture

There are many options when it comes to chicken housing on pasture. Lots of people opt for homemade chicken tractor–type structures that they can move to fresh grass every day. Giving chickens access to fresh grass means they'll have a great selection of bugs and whatever plants and seeds are growing in the grass. Even though the Cornish Cross is not a great forager, they still eat the occasional bug and seem to enjoy grass and weeds. Since the chickens poop a lot, moving the tractor

also helps distribute the manure and prevents overgrazing. Move your chickens every few days, and you'll find your grass and pasture will respond well to the added nitrogen in the manure.

Tips for Keeping Meat Birds Safe from Predators

Whatever type of housing you choose for your broilers, you will need to make sure it provides adequate shelter. Some protection from the sun is necessary, and you will want to make sure it has proper fencing, too, to prevent predator problems. Believe it or not, just about everything eats chickens: raccoons, skunks, foxes, coyotes, weasels, dogs, and bears. If your chickens are not in a covered area, you will also have to worry about aerial predators: eagles, owls, and hawks eat chickens, too.

To keep your meat birds safe from all their many predators, a covered chicken tractor is a great option. You will want to cover at least the bottom of it with woven chicken wire, not only to keep the chickens in, but also to keep the predators out. Another good tactic to keep out predators might be to surround it with one strand of electric fence wire. Not only can predators like raccoons reach in and snatch

your chickens, but hawks can grab them out of midair, and animals can try to dig under the tractor to grab them, so you need to cover all bases. You can find lots of plans for building your own chicken tractors online. If you'd like to free-range your chickens, electric poultry netting is often effective at keeping out four-legged predators but will not be effective against aerial predators. Livestock guardian companion animals can also be useful. We have a pair of Great Pyrenees dogs that do a wonderful job. Their nocturnal nature helps discourage predators at night when our chickens are most vulnerable.

FEEDING REQUIREMENTS

There are a couple of different approaches when it comes to feeding meat chickens. Some sources recommend a twelve-hour on/twelve-hour off feeding schedule. Giving the Cornish Cross birds a break from food is supposed to slow their growth a little bit and help prevent some of their common medical issues. Other people say that the twelve-hour on/twelve-hour off feeding schedule stresses broilers and they end up too hungry before the next feeding time. Because they are so hungry, they can claw at each other and cause injuries while they try to get to their food at the next feeding time. Many county extension offices recommend twenty-four-hour access to food as the most efficient way to grow broilers.

I let the feeders go empty on my Cornish Cross and feed them more along the twelve-hour on/twelve-hour off guidelines. Even with that feeding schedule, I still have the occasional medical problem with my Cornish Cross. It's also true that they clamber for the food when they see me coming at the next feed time. To prevent them from clawing at each other, I try to offer food in multiple locations so they can spread out to eat with less competition. I have not seen them injure each other due to frenzied, starving behavior. It's also a good idea to keep the food and the water separated from each other. This is a way to make sure the lazier broilers get at least a little bit of exercise throughout the day, which may help prevent health problems.

There are also different thoughts on what type of food to feed your broilers. Some people feed their meat birds the same amount of protein food for most of their lives. Other growers think that the chickens need differing protein content in their food as they grow. My approach is consistent with what most sources recommend:

Feeding Schedule
- Age 1 day to 3 weeks, feed broiler chicks a 21% to 22% protein chick starter.
- From 3 to 7 weeks, decrease protein to an 18% to 19% grower formula.
- From 8 weeks to processing, reduce protein to a 16% to 17% finisher formula.

It takes approximately 14 to 15 pounds of feed to feed one chick for 8 weeks. Chickens should convert food at a rate of 2.5 pounds of feed of live weight when they are processed between 8 and 9 weeks. At 8 weeks of age, the average Cornish Cross broiler will probably weigh about 6.5

pounds. Note that some broilers will be heavier, and some broilers may be lighter. Male broilers typically get bigger than female broilers. Since you can expect a dressed carcass weight of 70 to 75 percent of their live weight, a 6.5-pound chicken should equate to a 4- to 4.5-pound carcass for your freezer.

If you wait to process your broilers later (perhaps you are raising a slower-growing ranger or heritage breed), the food conversion weight goes up slightly to just under 6 pounds of feed per pound of live weight if processed around 10 to 11 weeks. Most rangers take at least 12 weeks to reach market weight, so this slower food conversion means they will be a more expensive broiler to raise.

Another Money-Saving Tip on Chicken Feed

Chicken feed is by far the largest day-to-day expense of raising chickens for meat, with organic or other specialty foods costing almost double what conventional chicken feed will cost. As suggested earlier, forming a food co-op and going in with other raisers on bulk orders can be one way to save a little money on chicken feed. Another way to save money is to feed your chickens some kitchen scraps, grow your own grain for your chickens, and let them free-range for bugs and other plant sources.

One of the best ways to save money on chicken feed is to reduce waste. Store it carefully so rodents or other birds cannot eat it. Keep it locked up tight in plastic or metal containers. We have had rodents chew through plastic containers to get to the chicken feed, however, so we switched to metal garbage cans. Do be aware that metal can sweat during hot summer days, so be mindful of feed getting wet. Five-gallon buckets with locking lids work well but, most importantly, keep your feed storage area clean, dry, and swept up to make it less appealing to rodents and bugs.

Treats for Meat Birds

There's a lot of emphasis on feeding meat chickens the right amount of protein to produce the best meat, and I provide my chickens with high-protein feed. However, I also supplement their diet by feeding them kitchen scraps, garden scraps, and treats, as well. While you certainly don't need to, broilers seem to enjoy their daily food scraps. We feed all our birds treats, and the broilers are no exception. Watermelon and watermelon rinds are a favorite, and the chickens do a great job of cleaning every little last bit of melon and leaving behind a paper-thin rind. Watermelon is a fun treat for both me and the chickens and is one of our favorites.

Keeping meat chickens on pasture gives them good access to nutrient-dense foods they love including clover, dandelion stems, flowers, roots, and seeds from chickweed and other types of "weeds" you may have mixed in with your grass. Even if you don't keep broilers on pasture, you can always bring the weeds inside to your chickens. My broilers also enjoy garden weeds and bugs that I collect and share, all the grubs I find in the garden, bugs that happen to live under feeders kept on grass, cabbage leaves, and tomato horn worms.

Kitchen scraps are great treats for broilers, and I give them our kitchen compost, including meat scraps. Chickens are omnivores, so it's okay to feed them meat scraps. Many homesteaders recommend feeding their broilers leftover raw milk, cheese, clabber, and whey. Feeding these high-protein foods to your broilers is especially useful if you have dairy animals on the homestead. Clabbering raw milk by leaving it out for several days to thicken creates a natural dose of probiotics, which is good for chickens. All these dairy products are high in protein, too, which your broilers need.

Black-oil sunflower seeds are another treat I offer my mixed flock. If you decide to give your birds sunflower seeds, it's important they are sunflower seeds for *birds*, not flavored or salted sunflower seeds that humans prefer. I pick up a fifty-pound bag of sunflower seeds in the animal feed aisle at the farm supply store to share with my mixed flock.

Food You Should Definitely Not Feed Your Meat Birds

While we feed our backyard birds a lot of food scraps, there are a few things you should avoid feeding them. It's not a good idea to feed your chickens tomato and other nightshade plant leaves or stems. Nightshade plants contain a toxin called solanine, which destroys red blood cells and can cause diarrhea and heart failure in your chickens. Save your chocolate for yourself and don't give it to your chickens, either, as it contains theobromine, a chemical that is toxic to birds. Finally, the caffeine in coffee grounds can be harmful to chickens, and moldy food waste is never a good food to feed any of your animals, including chickens. That said, I have found my chickens to be pretty discerning when it comes to the food I give them. If they want to eat it, they do. If they don't want to eat it, they don't. I have not had any problems due to feeding my chickens kitchen scraps.

Immune Boosters for Broilers

Taking a few easy steps can help improve your birds' immunity and potentially help them stay healthier. These tips will likely use items you already have around the house, and none them are expensive.

Add **apple cider vinegar** to their water. Adding apple cider vinegar (make sure to choose a variety with the mother, or *Mycoderma aceti*) to your chickens' water provides them with many benefits. Not only can it help boost their immune system, but it also helps maintain their digestive health and guards against bad bacteria. The other great thing about adding apple cider vinegar to their water is that it helps prevent algae growth in their water. If you've ever noticed a green buildup in your chicken waterers, you can prevent it with apple cider vinegar. For optimal benefits, add one tablespoon of apple cider vinegar to chicken water once or twice a week.

Garlic is a natural immune booster for chickens as well as humans. It helps repel bloodsucking mites, as they don't like the taste, and can improve respiratory health of chickens. Adding a small amount of crushed fresh garlic to the chickens' water or sprinkling a small amount of dried garlic on their food is an easy way to give their immune systems a boost. Please note that your chickens might not like the taste of garlic in the water unless you offer it to them from the very beginning. If you decide later on to give your chickens garlic, you might want to sprinkle the dried garlic on their food instead of putting crushed garlic in their water.

Blackstrap molasses has historically been used in animal feed for a variety of purposes including to encourage animals to drink, as a binder for food, and a natural source of iron, calcium, magnesium, and potassium. Research indicates that blackstrap molasses can produce heavier broilers that consume less feed at finishing. To supplement with blackstrap molasses, put a small amount (four ounces per gallon) in chicken water. Too much can cause diarrhea, so make sure you only use a small amount.[7]

HEAT STRESS

Summer heat is harder on chickens than the winter cold. Chickens can't sweat, so it's important to recognize signs of heat stress. You may notice your chickens spreading out their wings and panting when it's hot, as chickens lose heat through respiration. This is perfectly normal behavior during hot weather, but if they become lifeless, lethargic, or their combs start to discolor, you will want to cool them off as soon as you can. Cool them off by putting them in cool (not ice cold) water,

7 E. K. Ndelekwute, et al. "Effects of Administration of Molasses through Drinking Water on Growth and Conformation Parameters of Meat–Type Chicken" *African Journals Online*, vol. 6, no. 1, 2010. (https://www.ajol.info/index.php/apra/article/view/76141).

but don't submerge their heads. Then, keep them in a cooler spot until they are completely recovered and running around as normal.

Better than treating heat distress, though, is preventing it in the first place. To keep your chickens from overheating, make sure they have access to fresh, cold water during hot weather. Make sure that their waterers are full and accessible. It's a good idea to add ice cubes if you can. Double-check water throughout the day, as chickens drink a lot more during hot weather. You'll also want to make sure your chickens have access to shade. I check on my chickens more often during hot weather and move food and water as necessary to keep it in the shade. If you don't have shade in tall grass or around trees, make temporary shade with tarps to give your chickens access. Using water misters is another great way to keep your chickens cool. If you don't have misters, you can wet the ground with a hose to give your chickens a place to wet their feet. Burying a frozen bottle partway in the ground to give them a spot to rub on is another good idea. Finally, make sure your meat birds have plenty of ventilation. All coop windows should be open. If it is especially hot, you may want to run coop fans, as well.

Part III
Butchering

The butchering process might be the most daunting part of raising broilers. Many people tell me that if not for the butchering, they would be more open to the idea of raising their own chickens for meat. Not only do you need to humanely take an animal's life, but you also have to pluck it, remove the guts, and get it ready for the freezer. It's a lot of work, but it can also be very rewarding. Here's what you need to know about butchering chickens.

HOW TO KNOW WHEN BROILERS ARE READY TO BUTCHER

After you've been raising broilers a while, you will probably be able to eyeball your chickens and know if they're big enough. Until then, however, weighing them is a good way to determine if they are ready to be butchered. An easy way to weigh chickens is to use a digital fish scale and a bucket. Hang a bucket from a scale and zero it out. Then put a chicken in the bucket and see how much she weighs. Once you have the live weight, it's just a little simple math to estimate the processed weight. A processed chicken weighs about 70–75 percent of a live chicken's weight. You can calculate how much the chicken carcass will weigh after processing. Then, you'll know how much you'd like your live chicken to weigh before processing. I like my chickens to dress out somewhere around 5 pounds, so I like my chickens to weigh between 7 and 8 pounds at butchering time.

Live Weight versus Finished Weight

Live Chicken	At 70% Finished Weight	At 75% Finished Weight
5 pounds	3.5 pounds	3.75 pounds
6 pounds	4.2 pounds	4.5 pounds
7 pounds	4.9 pounds	5.25 pounds
8 pounds	5.6 pounds	6 pounds

DECIDING HOW TO BUTCHER

There are two options for processing your chickens: You can do it yourself, or you can take your chickens to a processing facility and pay someone to do it for you. Both methods have benefits and drawbacks that you will need to carefully consider. There is absolutely no shame in either choice, and it really is a very personal decision. Here's a quick rundown of both types of butchering.

Many people prefer to butcher their own chickens because they can control 100 percent of the process. If you intend to use the entire chicken, butchering at home might be the best option, as butchering a chicken yields more than simply the meat for cooking. You can use the all parts of the chicken when you butcher it yourself. The chicken necks and feet can be used to make a rich chicken stock. If you have dogs or pigs, you can feed them the heart, lungs, and kidneys. Chicken feathers and blood can enrich your compost pile. Your egg-laying chickens will love to peck over the intestines, and the liver can be turned into pate, if you are so inclined. Some of these uses might sound gross, but they are all things our ancestors would have done. Waste not, want not, after all.

If you choose to butcher your own chickens, however, be prepared for a long day. The better your supplies, the easier the process will be, so investing in a few quality processing items is a good idea: killing cones, knives, and a turkey fryer at the least. A tall table, garden hose,

outdoor sink, tarps, coolers, and perhaps a feather plucker will also come in very handy. You might be able to borrow equipment from friends or rent equipment from your local county agriculture extension office, and recruiting friends to help on butchering day is always a good idea. The day will go smoother with more hands; it's difficult to do the whole process on your own if you are butchering a number of chickens. Include equipment costs in your butchering decision so you can accurately compare the two types of butchering.

Taking Your Chickens to Be Butchered by Someone Else

While there are good reasons to butcher your chickens on your own, there are also good reasons for taking them to a processing facility. Processing facilities can be an effective use of your time and money. For $3 to $4 per chicken, an animal-processing locker will kill, clean, bag, chill, and freeze your chickens for you. They will dispose of all the blood, feathers, and chicken parts you may not want. They have all the right tools on hand, so you don't have to buy supplies, and they're quick, efficient, and the finished chickens are nicely plucked and store-quality packaged.

The challenging part of taking your chickens to a processing facility can be finding a place that does it. Many animal lockers no longer process chickens because it's not cost effective due to stringent government food safety regulations and minimal demand for butchering birds. With only a few places that offer chicken processing in most states, finding a facility within driving distance can be a real challenge. You will likely need to schedule your processing day in advance, and you might not be able to adjust that date if their schedule fills up or if your broilers grow faster or slower. You'll also need a way to transport the chickens to the processing facility.

I have processed my own chickens both ways and must admit that I like the convenience of taking chickens to be butchered at our local locker. I love the quality of the professionally butchered chickens and I appreciate not having to kill the chickens myself. While I intrinsically believe being able to butcher your own chickens is an important skill to have, I am happy and a bit relieved to pay professionals to do the butchering for me. Be sure to research your state's laws, as they all differ.

INVOLVING (OR NOT INVOLVING) YOUR CHILDREN

It's natural for your children to be interested and sometimes concerned when it comes to butchering chickens. After all, few families stop to consider where their food comes from, and even fewer families have eaten animals they have raised. While I think it is very important that children understand the life cycle and know where food comes from, there might be times when it is better not to involve your children in the butchering process. Some children are better able to handle the process than others, so I recommend that you think about each individual child. Maybe they can't be involved the first time, but you can recruit them to help on future butchering days. I always leave the decision to be involved up to my children.

If your children want to be involved, butchering day can be a very educational experience. Not only can they help with age-appropriate tasks, but they can learn about humane methods of butchering, and you may be able to work in a few science lessons, as well. It's fascinating to talk about the organs removed from chickens and their roles. Some children are very interested to see the hearts, lungs, and intestines and understand how they work. Other children might be confused or even frightened by the process, so I recommend you never force a child to participate on butchering day. It might be better in the long run for some children to be removed from the process until they are naturally curious or mature enough to accept the reality of the situation.

THE NIGHT BEFORE YOU BUTCHER

Withhold Feed

Whichever method of butchering you choose, it is important that you withhold feed from your broilers for at least eight and up to fourteen hours prior to butchering. Withholding feed allows your chickens to process all the food they have eaten and eliminate waste from their systems. It ensures that they will have a clean crop, an empty gizzard, and empty intestines during processing. Accidentally rupturing the crop, intestines, or gizzard can contaminate the meat with bacteria, so withholding feed is perhaps the most important step to take before butchering. While it is important to withhold feed, do not withhold water. Your chickens should continue to have access to clean, cool water.

Gather Chickens

It will be easier to butcher your chickens if you place them in area where they are easy to retrieve on the day of butchering. You can put them in cages or a fenced area using poultry netting. Any enclosed area that it's easy to get in and out of with a chicken will make the day go smoother. If your broilers free-range, it might be easier to catch them the night before butchering when they are roosting.

Gather Equipment

If you choose to butcher the chickens yourself, you will need to gather quite a few supplies. Some you may have on hand, and others you may need to special order online. Don't wait until the last minute to sort out equipment or you may not find what you need. Here is a list of supplies you will need if you butcher your broilers yourself.

- Sharp knives
- Knife sharpener
- Kitchen shears
- Killing cones
- Turkey fryer and thermometer
- Table
- Buckets or trash cans and heavy-duty garbage bags
- Gloves, leather and latex (optional)
- Poultry shrink freezer bags
- Drinking water/safe garden hose(s)
- Outdoor sink
- Tarps
- Coolers
- Bags of ice
- Stainless steel bowls for any chicken innards you want to keep
- Hand soap and kitchen towels
- Surface disinfectant spray
- Paper towels
- Feather plucker (optional)
- Waterproof boots and apron (optional)
- Camping shade canopy (optional)
- Snacks to keep your volunteer help happy (optional)

Set up the Butchering Stations

If you choose to butcher the chickens yourself, it's important to have a plan and to set up stations. If possible, set up your stations the day before you intend to butcher so you are ready to go bright and early in the morning. Choose a shady location that has access to water and electricity if you need to plug in any equipment. If you don't have a shady spot, a camping shade canopy might come in very handy. Walk through the stations before you start and make sure their placement makes sense. If movement from one station to the next seems awkward, move the stations around until the steps flow seamlessly. None of these stations needs to be fancy; functionality is key.

On butchering day, set up the following stations:

Killing Station

If you're using a killing cone, attach it to something sturdy. Some people prefer to have multiple killing cones to make the process go more quickly. Note that killing cones come in different sizes, so you'll want to consider what size you think your chickens will be when you buy the cones. We use a medium-sized, galvanized cone for large broilers and young turkeys up to ten pounds, and it works for both our small and large broilers. We attach the killing cone to a board and then  secure the board to a fence post. You'll want to put a bucket under the killing cone to catch the blood as the chicken bleeds, so consider that when you decide how to attach the cone. If the chicken is too big for the killing cone, you can also hang the chicken upside down by the feet to keep them calm during the process. A loop of twine and a tree branch work well.

Scalding Station

The scalding station should be a short walk from the killing area. Scalding is where you will dunk your broiler in hot water to make the feather-plucking process easier. A turkey fryer is a very useful tool to boil a broiler-sized pan of water. They work well outside, often burn readily accessible fuel, and also have multiple uses around the homestead (anybody like to freeze sweet corn?). It's handy to use the thermometer that comes with the turkey fryer to monitor water temperature, and you'll also want a pot of ice-cold water for dunking the chicken after scalding to stop the cooking process.

Feather-Plucking Station

If you're plucking the feathers by hand, we've found it helpful to pluck on a tarp over a picnic table or other tall surface, so you don't have to bend over too much during the plucking. Collecting the feathers on a tarp or in a bucket will make the cleanup easier at the end. You will either need to hold the broiler or hang it by its feet to pluck. If you're using a feather plucking machine, make sure you have a garden hose at the ready in this station as you'll need one to make the machine operate.

Plucking is time consuming and tedious. If you plan to raise a lot of broilers, it might make sense to invest in a mechanical chicken plucker, as they speed up the plucking process tremendously. You can find instructions and kits online to build your own, or you can buy a prebuilt one, as well. A quick Google search will yield a lot of information about chicken pluckers. I always recommend looking for used equipment and borrowing a plucker to try it out before buying.

Evisceration Station

The evisceration station is where you will cut off the head and feet and remove the organs. It's nice to have a tall table at this station. A sink or a garden hose to rinse the chickens is necessary. You'll also want knives, cutting boards, bowls, and buckets to catch body parts as you remove them from the chickens. If you want to wear latex gloves, this is where you will need them. We also have lots of kitchen towels and hand soap here, as well as disinfectant spray to clean the work surface between chickens. The towels came in handy to cover equipment to keep flies off as well as dry our hands after washing.

Chilling Station

You'll need some way to quickly cool your chicken after the evisceration is done. A cooler full of ice-cold water near the evisceration station is an easy way to keep your chickens cool.

HOW TO BUTCHER A CHICKEN

While you can learn a lot about butchering chickens from reading, it is really helpful to butcher with an experienced person the first time. If that's not possible, it's a good idea to find a few videos on the Internet. Watching a skilled butcher is a great supplement to reading about the processes and techniques.

The first step in the chicken-butchering process is catching a chicken. We try to be very gentle when we catch the chickens. It's helpful to corral them somewhere not too far from the butchering area so you can easily get them. Cradle their wings as you walk to the killing cone to keep them from flapping. It's important to keep the chickens as calm as possible during this process, as stress on animals at butchering time can affect the meat. Then, turn the chicken upside down and hold it by its feet. As the blood rushes to the chicken's head, it will calm down tremendously. I'm always surprised by how calm the chickens are during this process. It's a little tricky to get them upside down, but once you have them upside down, they calm down immediately.

Next, place the chicken in the cone, with the chicken's belly facing you. The goal is for the chicken's neck to stick out of the bottom, so you have easy access. You might have to adjust the chicken in the cone by reaching up into the cone and pulling the head down to expose the neck. Using a very sharp knife (don't even try with dull knives—they just won't work), sever the chicken's jugular vein and carotid artery by making two deep slices along both sides of its neck just under the jaw. Hold the chicken's head steady with one hand and pull back the feathers while making the cuts with the other hand. We recommend wearing a thick leather glove on the hand that's holding the neck to help protect your hand from the sharp knife.

The blood should flow out at a steady and consistent pace and the bleeding process should only take a few minutes. While the chicken is bleeding, its eyes may be open, and it might turn its head some. If it's not bleeding steadily, make a deeper cut. Through the bleeding-out process, the chickens are very calm. They don't thrash around that much until just as they die. When you see the chicken convulse, and the eyes quit moving, it's time to move on to the next step. The hardest part of the process is slicing the chicken's neck and watching it die. It

might be necessary for you to walk away for a couple of minutes after slicing the chicken. Just make sure that the chicken is bleeding steadily enough before you walk away. Everyone reacts differently to the actual killing. I find it to be the most emotional part of raising broilers. I prefer not to do this step myself, so my husband makes the cut.

Once the chicken is dead, it's time to scald it. Dunk the chicken in hot water and swirl it around for 30 seconds to 1 minute.

The water temperature is very important, because if it's too hot, it will start to cook the chicken. If it's too cold, it won't be effective at loosening the feathers. We've found the ideal temperature to be between 140°F and 150°F. Make sure the water covers all the feathers, so dunk it up to the feet feathers. However, be careful, because the water is hot and you don't want to burn yourself! Some people use hooks to facilitate the dunking. Once you can easily pull off a few leg feathers, your chicken should be ready to pluck. You may also want to dunk the chicken in a pot of ice-cold water to stop the cooking process.

If you scalded the chicken well, the feathers should come off easily. You can either hang the chicken by its feet or hold it over a bucket or tarp. Whichever method you prefer, start pulling off all the feathers. If you don't intend to use the head or butt area, you can leave those feathers. All other feathers need to come off, though. Be careful when you're pulling off the larger wing and tail feathers, as you'll rip the skin if you pull too many at once. You may want to pull off the bigger feathers one at a time. You'll also want to pull out any stray hairs or pinfeathers left behind. A dull pinning knife is helpful in removing the pinfeathers. Once you've finished plucking, rinse the chicken well with the garden hose.

After the chicken is plucked, it's time to cut off the head and feet and to eviscerate (remove the chicken's organs). If you intend to keep any of the chicken parts, you'll want several different containers for them. Stainless steel bowls filled with ice water are probably best. You'll also need a garbage can

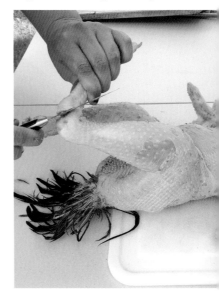

for the parts you aren't keeping. The first step is to cut off the head using kitchen shears. Then cut off the feet by slicing with a sharp knife in the valley between the two leg joints. Next, carefully cut off the oil gland just above the tail. You need to remove the oil gland, because it can affect the taste of your meat if it ruptures. You can either grab it and scoop down and out with the knife, or you can cut that whole back tail section off the chicken. Either way works depending on your preference. We found that cutting the whole tail off was easier when butchering a rooster with a lot of tail feathers.

Now it's time to clean up the inside of your chicken. We start at the head end by loosening the crop. The crop is a soft bag where the chicken's food is stored on its way to the gizzard for processing. If you didn't feed your birds the night before you butchered them, the crop should be empty. Some prefer it empty during butchering, but I honestly have an easier time dealing with the crop if there is a little bit of food in it. To loosen the crop, you will need to remove some skin from the neck. Then stick your hands down into the cavity and loosen the skin around the neck as you find the esophagus and trachea. The goal is to free the esophagus and trachea from the neck so they will come out when you remove the rest of the organs. If the crop is full, cut it out, but if it's empty, you can just loosen it. Let the esophagus, trachea, and crop hang from the body as you move on to the next step so they will pull out with the rest of the organs.

Now move to the bottom of the chicken. It's time to open up the chicken to remove the intestines and organs. Cut a small hole just above the vent, being careful not to rupture the intestines. Open the cavity by pulling the skin with both hands. You'll need to get your whole hand inside the cavity to pull out all the internal organs, so wear latex gloves during this step if you want. Once your hand is inside the cavity, you should be able to feel for the hard, round gizzard. Grab hold of the gizzard and pull, and the intestines, esophagus, trachea, and crop should come out, too. Hopefully the intestines came out intact and no fecal matter from the intestines leaked out inside your chicken. Another reason to withhold food prior to butchering is to help clear the intestines so you don't have to worry about

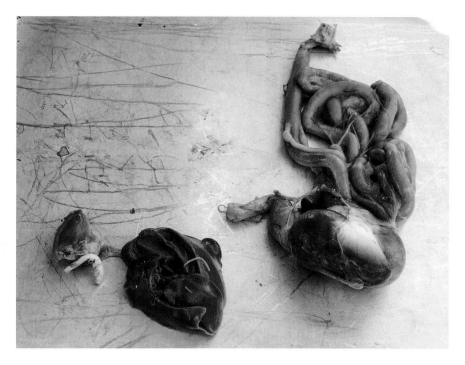

fecal matter. However, if fecal matter does spill, quickly run your hose to thoroughly clean out the chicken cavity and then disinfect your evisceration table.

The only ruptured organ that will ruin your chicken is the gall bladder. As you pull out the organs, you'll want to make sure you don't rupture the green gall bladder. You'll also want to make sure you get the heart and the liver, and that you clean the lungs off either side of the chicken's backbone, as well. The lungs are a spongy red color, and you can feel them quite well. It may take several swipes to get all the pieces out. Double-check that the trachea, esophagus, and crop came out with the rest of the internal organs. Once you cut off the tail portion, making sure to get the oil gland, you've survived evisceration! You'll get a lot faster at evisceration the more chickens you butcher, but the first time can be quite overwhelming.

The last thing to do is rinse your chicken out one more time and put it in an ice-water bath to chill for at least thirty minutes or overnight. After the chicken has chilled, dry it off and bag it, then put it back in the refrigerator for twenty-four hours before eating or moving to the freezer. After being butchered, the chicken will go through rigor mortis, which results in chemical changes to the meat. If you don't let the meat relax for at least twenty-four hours, it might be tough.

While the chickens are chilling, it's time to clean up and tear down. You can feed the feathers to your chickens, compost them, or bury them. We take the chicken parts we don't want and dump them on the edge of our property for scavengers to eat. You can also bury them a few feet in the ground.

Once everything is cleaned and stored, pat yourself on the back for a job well done! Taking responsibility for raising and processing your own food is a huge accomplishment, and you should be very proud.

The last thing you might want to do is reflect on the process. Did it go the way you wanted it to? Would a new piece of equipment or a minor change in process make the day go more smoothly? Take note of any changes you might want to make on future butchering days so you don't forget next time.

Checklist: The Evisceration Table

Here is a checklist of the steps described above that you will need to complete at the evisceration table after the chicken has been killed and plucked.

- Cut off the head with kitchen shears.
- Cut off the feet at the joint with a sharp knife.
- Cut off the oil gland and/or tail area near the vent with a sharp knife by angling out and downward.
- Cut the skin around the base of the neck to loosen the crop.
- Put your thumb through the neck between the esophagus, trachea, and crop.
- Pull the skin off the neck and remove the crop if it is full. You can leave it if it is empty.
- Cut off the esophagus and trachea tubes near the body.
- Cut off the neck close to the body by angling downward with your knife so no bones remain to break through storage bags.

- Make a small slit through the abdomen, careful not to rupture the intestines.
- Using both hands, open up the carcass.
- Put on latex gloves if you wish, and stick your hands all the way up the chicken carcass, loosening the organs.
- Grab the entrails and pull out the organs, intestines, and crop in one big swoop.
- If you are saving the heart and liver, put them in a bowl of ice water.
- Cut away the gizzard if you want to keep it. Cut open the gizzard longways and remove the contents and yellow lining. Rinse and set aside. Discard if you don't want to keep it.
- Put the intestines and any other pieces you don't intend to use in your discard bucket.
- Remove the red lungs from the rib cage in the upper cavity. This can be a little bit tricky, so keep scraping until you get both lungs.
- Rinse the chicken and put in the cooler to chill for two to three hours.
- Dry the chicken with paper towels. Put it in a shrink-wrap bag. Dunk in warm water to tighten the bag.
- Let rest overnight in the refrigerator.
- Put the chickens in the freezer and freeze.

Part IV
Cooking

One important aspect of raising your own chickens for meat is knowing how to cook the chickens. I know several people who go to all the work to raise and process meat birds, only to let the chicken die a slow death in the freezer because they don't know what to do with them. In this day, when most chicken recipes call for boneless, skinless chicken breasts or thighs, many people just don't know how to cook a whole chicken. We don't learn how to cut up a whole chicken, nor do we understand how to cook anything with bones and skin. However, a whole chicken is a versatile ingredient, and utilizing the whole chicken does not need to be intimidating!

It might take a slight perspective shift for some people to get over the distaste of working with skin and bones, but it is so worth it. By cooking a whole chicken and then chopping the meat, our family of six can easily get three full meals out of one chicken! Admittedly, that means that we eat less meat than your average family might, but by cooking a whole chicken and then using smaller portions of that meat as ingredients in larger recipes, we really stretch one chicken into multiple, delicious meals.

A note about chopped chicken: I see a lot of recipes that call for shredded chicken. My family doesn't enjoy shredded chicken, though. They think it's too stringy. To keep my family happily eating our home-raised chicken, I finally stopped shredding it and started chopping it instead. I find it to be easier to chop the chicken, and I will share a few tips for making the process easier in the recipes. But we don't stop with the chopped meat. The bones, organs, and other castoffs can be made into a delicious stock, too. Once you've turned your bones into stock, you can even feed them right back to your chickens. The whole chicken is a perfect low-waste, low-cost, real-food option for many families.

SAFETY CONCERNS FOR COOKING CHICKEN

There are some important safety concerns about cooking chicken that need to be mentioned. Salmonella, E. coli, listeria, and other harmful bacteria can be spread from uncooked chicken, so you'll need to take the following precautions to avoid salmonella and other harmful bacteria. Because cooking chicken thoroughly is the only way to kill salmonella, the USDA does not recommend raw chicken be rinsed and dried before cooking. Rinsing chicken spreads bacteria through your kitchen and has the potential to make people sick. If any recipe instructs you to rinse chicken, disregard that step. Rinsing chicken should not happen in your home!

The only way to kill harmful bacteria associated with chicken is cooking it thoroughly. While some recipes might lead you to believe that clear juices or loose joints indicate a fully cooked chicken, the only indicator that chicken is thoroughly cooked is internal temperature. You chicken is fully cooked when a meat thermometer inserted into the flesh of the thigh registers 165°F. Be careful that the thermometer does not hit the bone, or you'll get a false reading. Pink meat is not an indication that chicken is undercooked, so don't be alarmed if you see pink meat. As long as the internal temperature is 165°F, the chicken is safe to eat.

Thawing chicken properly is also an important safety consideration, and there are three approved methods for safely thawing chicken. The best way to safely thaw a whole chicken is to put it in the refrigerator and let it thaw slowly. Keep in mind that small, whole chickens will take around twenty-four hours to thaw, and large chickens will take even longer. So thinking ahead and pulling out frozen chicken a couple days before you need to cook it is key. It's also a really good idea to put your chicken in a baking dish so you don't end up with chicken juice all over your refrigerator. If you prefer, you can also thaw your chicken in cold water in the sink as long as its packaging is leak-proof. Thawing your chicken in a microwave is also a USDA-approved method of thawing, but it will need to be cooked right away after thawing. Never thaw your chicken on the counter. That's a recipe for breeding the harmful bacteria you want to avoid.

HOW TO CUT A WHOLE CHICKEN INTO EIGHT PIECES

While most of my recipes in this book use cooked, chopped chicken, there are a few recipes that require a cut-up chicken. Cutting up a whole chicken is more intimidating than it is difficult, but it's safer and easier if you invest in a good pair of poultry scissors. Here's how to cut up a whole chicken.

1. First, remove the backbone by cutting up along either side in the middle of the chicken.
2. Remove the leg quarters, simply cutting the skin that attaches them to the breast.
3. Cut the legs off the thighs.
4. Now cut off the wing tips and then wings. Add a piece of the breast meat to each wing to make it a more substantial piece.
5. Finally, cut the breast in half. If there are any pieces of backbone attached to the breast, cut them off as well.

Now you have turned one whole chicken into eight pieces. Save the backbone and any other pieces you may trim off for stock. I just put them in a large freezer bag and store in the freezer until I'm ready to make stock. Adding the backbone and other miscellaneous pieces to a chicken carcass is a great way to get a richer broth.

HOW TO SPATCHCOCK A CHICKEN

Spatchcocking, or flattening a whole chicken, is a great way to roast or grill chicken because it allows all the pieces to cook uniformly. To spatchcock a whole chicken:

1. Cut out the backbone with poultry scissors by slicing up both sides on the back of the chicken.
2. Remove the wing tips if you want.
3. Flatten the chicken with your hands.
4. Again, reserve the backbone and any other pieces you may cut off for stock.

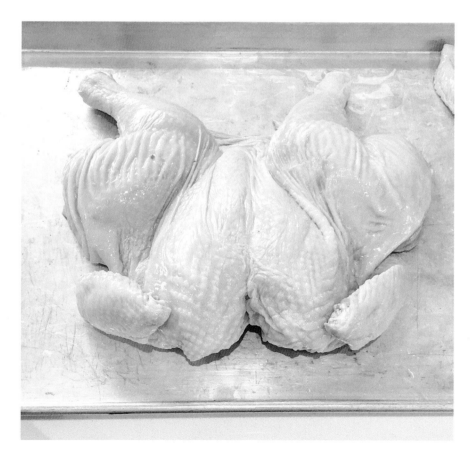

30 FARM-TO-TABLE RECIPES

Versatile Cooked Chicken (3 Ways)

Precooked chopped chicken is my mainstay for so many recipes. Think of it as a blank slate for just about any chicken recipe. Here are three easy ways to cook a whole chicken to use in other recipes: in an electric pressure cooker, a slow cooker, or in a stockpot on the stove. If you prefer shredded chicken to chopped chicken, go ahead and shred it!

Serves: varies depending on the size of the chicken
Prep time: 15 minutes
Cooking time: varies based on size of chicken and cooking method

Ingredients

1 whole chicken
1 onion, skin on, halved or quartered
2 carrots, peel on, cut into large chunks
2 garlic cloves, skin removed, halved
2 stalks celery, leaves are okay
1 tablespoon Rotisserie Spice Blend (page 129), or just salt and
  pepper if you prefer

Electric Pressure Cooker Directions

Combine all ingredients in the electric pressure cooker inner pot. Cook on manual high, using time chart below for timing based on chicken size. Naturally release pressure for 15 minutes, then manually release remaining pressure. Let chicken cool for 10 to 15 minutes, then check that the internal temperature has reached 165°F. Cook longer if need be, otherwise strain out and discard vegetables, reserving liquid for broth. When the chicken is cool enough to handle, cut or pick off the meat and coarsely chop or shred. Reserve the bones for stock.

Continued on page 78.

Size of Chicken	Fresh (thawed)	Frozen
3 pounds	20 minutes	39 minutes
3½ pounds	23 minutes	46 minutes
4 pounds	25 minutes	52 minutes
4½ pounds	28 minutes	59 minutes
5 pounds	33 minutes	65 minutes

Slow-Cooker Directions

Combine all ingredients in a 6-quart slow cooker. Cook on high for one hour. Reduce heat to low, and continue cooking for 4 to 5 hours or until the internal temperature has reached 165°F. When the chicken is cool enough to handle, cut or pick off the meat and coarsely chop or shred. Reserve the bones for stock.

Stockpot Directions

Combine all ingredients in a large stockpot. Add water to cover the chicken. Bring to boil over medium-high heat, then reduce temperature to low and simmer until chicken recaches an internal temperature of 165°F, about 60 to 90 minutes. When the chicken is cool enough to handle, cut or pick off the meat and coarsely chop or shred. Reserve the bones for stock.

Homemade Chicken Stock (3 Ways)

There are several ways to make homemade chicken stock: on the stove, in a slow cooker, or in an electric pressure cooker. My personal favorite is the electric pressure cooker. To get a rich bone broth, however, you might want to cook for longer (up to twenty-four hours) on the stove top or in the slow cooker. Here are three different ways to make a rich, delicious stock.

Serves: varies depending on size of pot
Prep time: 10 minutes
Cooking time: varies based on method used

Ingredients
1–2 chicken carcasses, meat removed
3 tablespoons apple cider vinegar
1 large carrot, chopped into large pieces
1 large onion, halved or quartered (leave the skin on if desired)
2 celery stalks, chopped into large pieces
1 head garlic, halved crossways
2 teaspoons salt
1 teaspoon whole peppercorns
Filtered water to fill the pot

Stove Top Directions
Put the meat carcass(es) in a large pot and fill the pot with filtered water. If you have extra backbones or other chicken pieces, add them to the pot, too. Add the apple cider vinegar and bring the pot to a gentle boil. Skim off any foam that rises to the surface.

Then add the chopped vegetables and spices and cover the pot. Continue boiling gently for 4 to 24 hours. Let broth cool to room temperature and strain bones and vegetables from the stock.

Continued on page 81.

Slow-Cooker Directions

Combine all ingredients in the slow cooker. Cook on low overnight, or for 8 to 24 hours. Let broth cool to room temperature and strain bones and vegetables from the stock.

Electric Pressure Cooker Directions

Combine all ingredients in electric pressure cooker. Using the sauté function, bring liquid to a boil and skim off any foam that rises to the surface. Lock lid and cook on manual high for 40 minutes. Manually release pressure. Let broth cool to room temperature and strain bones and vegetables from the stock.

Note: This stock will keep in the refrigerator for three to four days or for six to twelve months in a freezer.

Caramelized Onion, Bacon, Spinach & Chicken Quiche

If you need an impressive and delicious brunch recipe, here it is! Featuring caramelized onions, bacon, and spinach in addition to left-over chicken, this is one of my favorite quiches to make.

Serves: 6
Prep time: 25 minutes
Cooking time: 45 minutes–1 hour

Ingredients

1 prepared piecrust (page 126)
4 thick slices bacon
1 large onion, chopped into thin strips
1 cup chopped greens (spinach, kale, etc.)
2 cloves garlic, crushed
1 tablespoon fresh chopped thyme, or 1 teaspoon dried

½ teaspoon salt
½ teaspoon pepper
6 large eggs
1 cup shredded cheddar cheese
¾ cup heavy whipping cream or half-and-half if you prefer
¾–1 cup chopped cooked chicken

Directions

Preheat oven to 350°F. Prepare a 9-inch pie pan with the bottom piecrust. Poke a few holes in the crust with a fork so that it won't poof up.

Chop the bacon into small lardon pieces and cook over medium-high heat. Remove bacon from pan and drain all bacon grease, reserving 2 tablespoons.

Cook the onion in the bacon grease until caramelized, about 15 to 20 minutes. When the onions have about 5 minutes left to cook, add the greens, the crushed garlic, thyme, salt, and pepper. Stir well and continue cooking for 5 minutes. Then turn off heat and set aside.

In a medium mixing bowl, whisk eggs, then add cheese and whipping cream. Stir well to combine. Add the caramelized onion/vegetable mix, the chicken, and the bacon and stir well to combine.

Put the mix in the piecrust and bake for 45 minutes to 1 hour, or until eggs are set and a toothpick comes out clean.

Remove from oven, let rest for 5 to 10 minutes, cut, and serve.

Greek Chicken Mini Frittatas

Delicious to serve for brunch or great for a make-ahead breakfast that you can freeze and reheat for busy days, these Greek Chicken Mini Frittatas are bursting with flavor and perfect for people who love Greek flavors. This is a great way to use up any leftover chicken when you only have a small amount!

Serves: 6 (2 muffins each)
Prep time: 15 minutes
Cooking time: 25 minutes

Ingredients

12 eggs
1 cup chopped spinach
½ cup halved cherry tomatoes
¼ cup chopped kalamata olives
½ cup chopped cooked chicken
½ cup feta cheese
1 teaspoon Greek Seasoning Spice Blend (page 129)
½ teaspoon salt

Directions

Preheat oven to 350°F. Prepare a 12-cup muffin pan by spraying with nonstick spray or lightly coating with olive oil or butter.

In a large bowl, whisk eggs, then add vegetables, chicken, cheese, and spices.

Carefully fill the muffin wells about ¾ of the way.

Place muffin tin on baking sheet, and bake for 25 minutes or until the muffins are golden brown and the eggs are set.

Remove from oven and let rest for 5 minutes. Then carefully run a butter knife around each muffin to help release them from the pan. To freeze muffins, cool completely, then flash freeze on a cookie sheet for one hour. Next, bag them in a freezer bag and store for up to 3 months. To reheat, defrost frozen muffins in the microwave until warm, 2 to 3 minutes.

Enjoy!

Hearty Chicken Noodle Soup

Ditch the canned soup and make your own! This easy, wholesome, and hearty soup is the perfect use for leftover chicken and homemade stock. Chicken noodle soup isn't just for sick people anymore! Serve with a fresh salad and crusty bread for a quick weeknight meal.

Serves: 6
Prep time: 15 minutes
Cook time: 30 minutes

Ingredients

1 tablespoon avocado oil	2 celery ribs, sliced ½-inch thick
1 onion, finely chopped	1 tablespoon tamari sauce
3 large garlic cloves, minced	3 cups chopped cooked chicken
1 teaspoon fresh thyme minced, or ¼ teaspoon dried	4 ounces dried wide egg noodles
8 cups Homemade Chicken Stock (page 79)	¼ cup fresh parsley, minced
	Salt, to taste
4 carrots, peeled and sliced ½-inch thick	Pepper, to taste
	1 tablespoon Rotisserie Spice Blend (see page 129), optional

Directions

Heat oil in a large stockpot over medium-high heat and sauté chopped onion until softened, about 5 minutes. Then add the garlic and thyme and cook about 1 minute longer.

Add stock, carrots, celery, and tamari sauce, scraping any browned bits off the bottom of the pot. Bring to a gentle boil, reduce heat, and simmer for 20 minutes.

Add cooked chicken, turn up the heat, and bring the soup back to a boil. Add noodles and cook according to package directions.

Turn off heat, stir in parsley, add salt and pepper to taste, and Rotisserie Spice Blend if desired.

Serve immediately and enjoy.

Creamy Chicken Tomato Tortellini Soup

A hearty, rich, and creamy soup. Perfect for cool fall evenings or busy nights when you need a quick meal on the table.

Serves: 6
Prep Time: 10 minutes
Cooking Time: 20 minutes (dependent on tortellini)

Ingredients

2 teaspoons avocado oil
1 medium yellow onion, finely chopped
3 cloves garlic, minced
4 cups chicken stock (page 79)
1 (14.5-ounce can) petite diced tomatoes
1 (14.5-ounce can) tomato sauce
1 tablespoon dried basil
2 cups dried tortellini, or frozen if you prefer
3 cups chopped cooked chicken
3 cups packed chopped spinach
½ cup freshly grated Parmesan cheese
1 cup half-and-half

Directions

Heat oil in a large pot over medium-high heat. Then add onion and cook for 2 to 3 minutes until fragrant. Stir in the garlic and cook for 2 to 3 more minutes, or until the onions are translucent.

Add the chicken stock, tomatoes, tomato sauce, and basil, and bring mixture to a boil.

Add the dried tortellini and cook per instructions on the package. When the tortellini have 3 minutes left to cook, add the chicken and spinach. Once the tortellini are finished, turn the temperature to low and add the Parmesan cheese and half-and-half. Add salt and pepper to taste and enjoy.

Creamy Chicken Stew

When you're in the mood for creamy stew, there's not much better than this one! Fresh garlic, kefir, and bone broth make this an immune-boosting recipe, too. This hearty stew is another one of our family's favorite soup meals. Serve with crusty bread and a fresh salad.

Serves: 4
Prep Time: 15 minutes
Cooking Time: 20 minutes

Ingredients

1 tablespoon avocado oil
1 medium onion, chopped
3 cloves garlic, crushed
3 medium carrots, chopped
2 large ribs celery, chopped
4 cups chicken bone broth
1 teaspoon dried thyme
1 cup cubed, peeled potatoes
2 cups chopped cooked chicken
⅔ cup fresh or frozen peas
1 teaspoon salt
½ teaspoon finely ground black pepper
4 tablespoons unsalted butter
¼ cup all-purpose flour
2 egg yolks
½ cup kefir or half-and-half
Fresh parsley for garnish

Directions

Heat oil in large pot over medium-high heat and add onion, garlic, carrots, and celery. Cook and stir until the vegetables are tender, about 6 minutes.

Add broth, thyme, and potatoes, cover, and bring to a gentle simmer over high heat. Simmer until the potatoes are cooked, about 15 minutes.

Turn the heat to low, and add the cooked chicken, peas, salt, and pepper.

Over medium heat melt butter in a small saucepan. Add flour and stir to form a smooth paste.

Gradually stir the paste into the stew and simmer over low heat until the stew thickens, 3 to 4 minutes.

Mix together the egg yolks and the kefir in a small bowl. Add ¼ cup hot stew to the mix and then whisk the egg mixture back into the stew. Cook on low for 1 more minute.

Garnish with parsley and enjoy!

Chopped Chicken Salad

Using lots of fresh vegetables from the garden, leftover chicken, and a quick homemade dressing, this delicious salad is perfect for hot days when you don't want to heat up the kitchen.

Serves: 8
Prep Times: 15 minutes

Ingredients

6 slices bacon
1 large tomato
1 medium cucumber
1 avocado
1 head romaine lettuce, washed and chopped into bite-size pieces (about 8 cups)
2 cups cooked leftover rotisserie chicken, cut into bite-size pieces
½ cup feta cheese
¼ cup sunflower seeds
Homemade Ranch Dressing (page 94)

Directions

Slice the bacon into bite-size pieces and cook until crispy. Remove from grease to drain. Save the bacon grease to use in other recipes, if you want.

Chop the tomato, cucumber, and avocado into bite-size pieces and add to the bowl with the lettuce.

Add the chicken, feta cheese, and sunflower seeds.

Finally, add dressing to taste and toss well to combine.

Note:

This salad makes a really great meal, but it won't keep long after mixing everything together. If you think you'll have leftovers, it's better to only add dressing to the portion you think you'll eat. Because the acid in the dressing will make the lettuce soggy once the two are mixed together, keep them separate until you intend to eat.

Homemade Ranch Dressing

This delicious salad dressing is fresh, healthy remake of a store-bought favorite. Made with Greek yogurt and fresh herbs, this homemade ranch is a huge hit at our house for salads, dipping, fries, and more.

Serves: 8
Prep Times: 10 minutes

Ingredients

1 tablespoon fresh parsley (or 1 teaspoon dried)
1 tablespoon fresh chives (or 1 teaspoon dried)
1 tablespoon fresh dill (or 1 teaspoon dried)
1 garlic clove (or ½ teaspoon garlic powder)
1 cup whole milk Greek yogurt
1 teaspoon dried onion powder
½ teaspoon salt
½ teaspoon black pepper
1 teaspoon Dijon mustard
1 teaspoon fresh lemon juice
3 tablespoons whole milk, or to desired consistency

Directions

If using a food processor, pulse the fresh herbs and garlic several times until well chopped. Add the remaining ingredients except for milk and pulse to combine. A slightly mounded palmful of unchopped fresh herbs should equal about 1 tablespoon chopped. I eyeball it.

Add 1 tablespoon of milk at a time, until you reach the desired consistency. I added about 3 tablespoons for a runnier salad dressing but would omit milk for a creamier dip.

Store in the fridge for up to one week.

Roasted Tomato & Blueberry Salad with Chicken

This delicious salad takes a little bit of time to prepare, but the result is worth the effort! You can speed up the process by candying the walnuts and roasting the tomatoes and blueberries in the morning.

Serves: 6
Prep time: 25 minutes
Cook time: 45 to 60 minutes

Ingredients

1 cup cherry tomatoes, halved
1 cup fresh blueberries
¼ cup extra-virgin olive oil
¼ cup balsamic vinegar
1 cup chopped walnuts
¼ cup sugar
10 cups spring mix salad or lettuce
2 cups cooked leftover chicken
½ cup herbed feta cheese, or
   herbed goat cheese

Vinaigrette Dressing

2 tablespoons oil/vinegar left on
   the pan after roasting tomatoes
   and blueberries
1 tablespoon extra-virgin olive oil
½ tablespoon balsamic vinegar
1 clove garlic, minced
1 tablespoon raw honey
1 teaspoon Dijon mustard

Directions

Preheat oven to 425°F. Put halved tomatoes and blueberries on a roasting sheet and drizzle olive oil and vinegar over them, stirring to coat them all. Put them in the oven and roast 25 minutes. Remove to cool. Reserve 2 tablespoons liquid from the baking sheet.

Meanwhile, combine walnuts and sugar in a medium-sized skillet over medium-high heat. When all the sugar is melted, turn the heat off, and remove the nuts to a piece of aluminum foil to cool.

Combine all vinaigrette ingredients in a screw-top jar and shake to combine.

Put washed lettuce in a large bowl, and add chicken, feta cheese, caramelized walnuts, and roasted tomatoes and blueberries. Toss with dressing and serve immediately.

Rotisserie Chicken (2 Ways)

Why buy a rotisserie chicken when you can make your own without any special rotisserie cookers? You can get a very similar chicken using your slow cooker or electric pressure cooker and the flavor of your homegrown chicken will be out of this world!

Serves: 8
Prep time: 40 minutes
Cooking time: 45 to 60 minutes

Ingredients
4-pound whole chicken, fresh or frozen
2½ tablespoons Rotisserie Spice Blend (page 129)
2 tablespoons avocado oil

Slow-Cooker Directions
Tie the legs together with cooking twine. Mix the spices with the oil and brush the spices all over the chicken.

Cook the chicken on high heat for 1 hour, then reduce heat to low and continue cooking for 4 to 5 hours, or until the internal temperature of the chicken has reached 165°F.

Carefully remove the chicken to a cookie sheet. To develop a crispier skin, broil chicken in the oven for 2 to 3 minutes, or until nicely browned. Serve immediately and enjoy.

Electric Pressure Cooker Directions
Tie the legs together with cooking twine. Mix the spices with the oil and brush the spices all over the chicken.

Put the chicken in the pressure cooker on manual high for 25 minutes. Allow the pressure to release naturally for 15 minutes, and then quick release any remaining pressure. Check to make sure the chicken has reached 165°F. If not, re-seal and cook an additional 5 to 10 minutes. Carefully remove the chicken to a cookie sheet.

To develop a crispier skin, broil chicken in the oven for 2 to 3 minutes, or until nicely browned. Serve immediately and enjoy.

Coq au Vin

A classic French dish, *Coq au Vin* literally means "rooster with wine." This traditional French recipe is a great way to cook a tough old rooster. Featuring a whole, cut chicken, garden-fresh vegetables and herbs, and a lot of wine, it's a wonderful meal to serve company or your family when you have a little extra time on your hands. Here's my take on Coq au Vin.

Serves: 8
Prep time: 40 minutes
Cooking time: 45 to 60 minutes

Ingredients

4 thick slices of bacon sliced into small pieces
Salt, to taste
Pepper, to taste
1 whole chicken, cut into 8 pieces (page 72)
2 yellow onions, chopped into large pieces
2 medium carrots cut into 1-inch pieces
3 cloves garlic, minced
8 ounces button mushrooms, halved or quartered

¼ cup cognac or brandy
3 tablespoons all-purpose flour
1 tablespoon chopped fresh thyme, or 1 teaspoon dried
1 teaspoon dried marjoram
1½ cups burgundy wine
½ cup Homemade Chicken Stock (page 79)
2 bay leaves
Mashed potatoes or egg noodles, for serving

Directions

Preheat the oven to 325°F. Cook bacon pieces in a Dutch oven over medium-high heat then remove to a plate.

Salt and pepper the chicken pieces, and working in batches, sauté in the hot bacon grease for 4 to 5 minutes per side, until the chicken is nicely browned.

Continued on page 100.

Once the chicken is browned, add the onions and carrots to the bacon grease. Sauté for 4 to 5 minutes, or until the onion are nice and fragrant, and then add the garlic and mushrooms and sauté for 2 to 3 more minutes.

Now add the cognac, flour, thyme, and marjoram, and stir until bubbly. Place the chicken back in the pot on top of the vegetables and pour the wine and chicken broth over the chicken. Add the bay leaves, and bake, covered, for 45 minutes to 1 hour, or until the chicken is thoroughly cooked to 165°F.

Remove bay leaves before serving. Serve over egg noodles or mashed potatoes and enjoy!

Grilled Beer Can Chicken

Developed by Nicky Omohundro from the popular family travel blog *Little Family Adventure*, this delicious recipe is a great way to use your whole chickens. You won't be able to get enough of this recipe, which is seasoned with a homemade coffee brown sugar rub.

Serves: 8
Prep time: 10 minutes
Cook time: 60 to 75 minutes

Ingredients

2 tablespoons ground coffee
2 tablespoons brown sugar
1 tablespoon sea salt
1 tablespoon smoked paprika
2 teaspoons ground black pepper
1 teaspoon onion powder
1 teaspoon garlic powder

1 teaspoon cumin
1 teaspoon dried sage
1 teaspoon dried marjoram
Pinch cayenne pepper (optional)
2 whole chickens
2 cans beer

Directions

Start the grill and preheat to 375°F.

Mix ground coffee and all spices together, then rub the exterior of the chickens with coffee brown sugar rub. Open 2 cans of beer and dump out a little (or take a few big sips out of them).

Place each chicken on top of a can, inserting the can into the chicken's cavity. Place chicken upright on a section of the grill, NOT over coals or burners.

Close lid and cook for 60 to 75 minutes, or until the internal temperature reaches 165°F. Remove from grill and let stand for 10 minutes. **Note:** Cans will be hot! Carefully remove the cans.

Cut up chicken and serve.

Roasted Rosemary Lemon Spatchcock Chicken

Roasted Rosemary Lemon Spatchcock Chicken is delicious main course, perfect for weekend family dinners or even to serve to company. With a crispy skin and a fresh flavor, this recipe is sure to become a quick favorite recipe.

Serves: 4–6
Prep time: 40 minutes
Cooking time: 45 to 60 minutes

Ingredients
1 lemon
2 tablespoons extra-virgin olive oil
2 tablespoons chopped fresh rosemary
1 clove garlic, minced
1 teaspoon salt
½ teaspoon pepper
1 spatchcocked chicken

Directions
Preheat oven to 475°F. Cut lemon in half and set aside.

Combine olive oil and all other ingredients except lemon in a small bowl and whisk to combine.

Brush this mixture on top of the chicken and under the skin.

Place chicken on a roasting pan, add the lemon, and bake for 35 to 40 minutes or until the internal temperature of the chicken reaches 165°F.

Remove chicken from oven, squeeze lemon over chicken. Let rest for 5 to 10 minutes, then cut and serve.

Greek Grilled Spatchcock Chicken

This Greek Grilled Spatchcock Chicken will be the hit of your summer BBQ. The marinade is great for cut chicken pieces, as well, so keep the recipe handy whenever you are grilling chicken!

Serves: 4
Prep time: 20 minutes active + longer for marinating
Cooking time: 45 to 60 minutes

Ingredients

1 cup whole milk plain yogurt
2 tablespoons extra-virgin olive oil
4 cloves garlic, minced
Juice of ½ a lemon, plus zest if you want
2½ tablespoons Greek Seasoning Spice Blend (page 129)
1 spatchcocked whole chicken

Directions

Combine all marinade ingredients in a glass bowl and stir well. Add chicken and cover with marinade. Refrigerate until ready to grill, for at least 30 minutes, but up to 24 hours.

Heat the grill to medium-high and grill chicken breast-side down for 10 to 15 minutes, or until nicely charred. Cover the grill but leave the vent open. Turn down the grill to low, flip the chicken, and grill for 30 more minutes. Then flip the chicken one more time, and cook until the internal temperature of the chicken reaches 165°F.

Chicken Alfredo with Kale & Broccoli

My mother makes chicken alfredo for her grandkids, and it is one of their favorite meals. They always ask me to make this rich recipe for them. I've made it a little bit healthier by adding kale and broccoli from my garden, but my kids prefer it without the vegetables. Feel free to omit, if you prefer.

Serves: 6
Prep time: 10 minutes
Cook time: 20 minutes

Ingredients
½ cup unsalted butter
½–1 cup kale, washed and cut into bite-size pieces
2 cloves garlic, minced
2 cups heavy cream, or half-and-half for a lighter dish
¼ teaspoon salt
¼ teaspoon white pepper
1½ cups grated fresh Parmesan cheese
2 cups chopped cooked chicken
1 cup steamed broccoli
1 pound cooked noodles of choice

Directions
Melt butter in a medium-sized saucepan over medium heat, then add kale, garlic, cream, salt, and pepper.

Bring to a low boil, then reduce heat to medium-low. Simmer for approximately 8 minutes.

Remove from heat and then slowly add the cheese. Return to low heat and cook for 2 to 3 more minutes, stirring constantly.

Add chicken and broccoli, and stir to combine.

Serve over cooked noodles, and enjoy!

Blackened Chicken Tacos with Slaw & Lime Crema

We love blackened tacos! Using leftover chicken and premade seasoning mixes, this recipe is really quick to make if you can prepare the slaw and crema earlier in the day.

Serves: 6
Prep time: 25 minutes

Ingredients
3 cups chopped cooked chicken
2 tablespoons Spicy Blackened Seasoning Blend (page 127)
½ cup Homemade Chicken Stock (page 79)

Red Cabbage Slaw
½ head small red cabbage, finely chopped
¼ cup fresh cilantro, chopped
¾ cup thinly sliced red onion
Finely chopped jalapeño, to taste

1½ tablespoons fresh lime juice
1 tablespoon extra-virgin olive oil
1 tablespoon apple cider vinegar
½–1 teaspoon salt

Lime Crema
½ cup sour cream or whole milk Greek yogurt
2 teaspoon fresh lime juice
½ teaspoon finely grated lime zest
Corn tortillas

Directions
Mix cooked chicken, Spicy Blackened Seasoning Blend, and chicken stock in medium saucepan and heat until warm. Set aside.

To make the slaw, toss cabbage, cilantro, onion, and jalapeño in a medium-sized bowl. Whisk lime juice, olive oil, vinegar, and salt, then pour over cabbage mix. Chill in refrigerator for at least 30 minutes to allow the flavors to meld.

To make the lime crema, stir sour cream, lime juice, and lime zest until well combined.

To serve, heat corn tortillas and top with chicken, slaw, and crema.

Chicken Burrito Bowls

This is one of my family's favorite uses for leftovers. I always try to make extra rice, beans, and chicken to have on hand for quick Chicken Burrito Bowls. If we don't have rice, we'll also turn these into taco salads.

Serves: 6
Prep time: 10 minutes

Ingredients

3 cups chopped cooked chicken
2–3 tablespoons Taco Seasoning Blend (page 127)
¼ cup Homemade Chicken Stock (page 79)
3 cups cooked rice
2 cups cooked black or pinto beans

Toppings of choice:
Shredded lettuce
Diced tomatoes
Chopped olives
Shredded cheddar cheese
Sour cream
Salsa
Guacamole

Directions

Heat chicken in a medium saucepan over medium heat with taco seasoning and chicken stock, for 5 minutes or until heated through.

To assemble your chicken burrito bowls, spoon rice, beans, and chicken in large bowls. Add toppings of choice.

Chicken Fajita Sheet-Pan Dinner

While many of my recipes utilize cooked chicken, raw chicken is really preferable in this one. I cut as many large chunks of chicken as I can off a whole chicken and then slice into strips. Make sure to reserve the carcass for soup because there will still be lots of chicken left on it. Your family will love this no-fuss chicken fajita cooking method. Made on one large sheet pan in the oven, the flavors are amazing, and cleanup is a breeze.

Serves: 6
Prep time: 10 minutes
Cooking time: 20 to 25 minutes

Ingredients

3 bell peppers (red, orange, green or whatever you have on hand), sliced in strips
1 onion, sliced in strips
3 cups thinly sliced chicken strips
2 tablespoons avocado oil
2 cloves garlic, minced
2 tablespoons Taco Seasoning Blend (page 127)
Juice of 1 lime
¼ cup fresh cilantro, chopped

Directions

Preheat oven to 425°F. Spray a sheet pan with cooking spray and place vegetables and chicken on tray. Mix oil, garlic, and taco seasoning in a small bowl and pour over chicken and vegetables. Bake for 20 to 25 minutes until chicken is thoroughly cooked.

Squeeze lime and sprinkle with cilantro to taste after cooking. Enjoy!

Roast Chicken & Vegetables Oreganato

This delicious roast meal is quick to prepare and fitting for company or a Sunday dinner. Featuring fresh herbs, farm-fresh vegetables, and home-raised chicken, it's a one-dish meal your whole family will enjoy.

Serves: 6
Prep time: 15 minutes
Cook time: 1 hour

Ingredients

4 medium potatoes, cut into medium-sized chunks

1 cup cherry tomatoes, halved

1 small broccoli, cut into bite-sized pieces (optional)

1 large onion, cut into medium-sized chunks

1 whole chicken, cut into 8 pieces (see page 72)

⅓ cup extra-virgin olive oil

2 garlic cloves, minced

¼–½ cup fresh oregano, coarsely chopped

1 teaspoon fresh lemon juice

½ teaspoon salt

½ teaspoon freshly ground pepper

Directions

Preheat oven to 425°F. Place chopped vegetables in a 9 x 13–inch baking dish and top with the chicken skin-side up.

Combine olive oil, garlic, oregano, lemon juice, salt, and pepper in a small bowl to make a marinade. Whisk to combine. Pour marinade over chicken and vegetables.

Bake for 1 hour until the chicken juices run clear and/or a meat thermometer measures 165°F.

Note: If you wish to marinate the chicken before baking, put chicken pieces in a large glass bowl and pour marinade over to coat the chicken. Marinate in the refrigerator for at least 30 minutes, but up to 24 hours. When you're ready to bake, pour the chicken and marinade over the vegetables in the 9 x 13–inch baking dish and follow baking instructions above.

Chicken Taco Zucchini Boats

The perfect mid-to-late-summer meal when the garden is overflowing with zucchini! These flavorful chicken taco zucchini boats are delicious and filling.

Serves: 6
Prep time: 20 minutes
Cook time: 20 to 25 minutes

Ingredients

4 medium zucchini
1 tablespoon avocado oil
½ medium pepper, finely diced
½ medium onion, finely chopped
2 cups chopped cooked chicken
2 cups cooked corn (great use for any leftover ears of corn on the cob you may have!)
2 cups cooked black beans (about 1 can)
2 tablespoons Taco Seasoning Blend (page 127)
1 cup shredded cheddar cheese

Directions

Preheat oven to 400°F.

Slice zucchini in half lengthwise and scoop out the insides. Reserve half the scooped zucchini and save the rest for another dish. Place the zucchini on a baking sheet.

In a medium-sized pan over medium-high heat, warm the avocado oil, then sauté the pepper and onion for 2 to 3 minutes. Add in the reserved scooped zucchini and sauté several more minutes until the vegetables are tender and fragrant.

In a large bowl, mix together the chicken, corn, beans, taco seasoning, and sautéed vegetables.

Fill the zucchini boat halves with the mixture and top with shredded cheese.

Bake for 20 to 25 minutes, or until the zucchini has reached the desired tenderness. I like my zucchini soft, so I bake for 25 minutes.

Enjoy!

Chicken Tikka Masala

This flavorful dish is one of our family's favorites. Using cooked, chopped chicken means it is quick and easy to prepare. Perfect for cold nights or any night you need dinner on the table in under 30 minutes!

Serves: 6
Prep time: 10 minutes
Cook time: 20 minutes

Ingredients

2 tablespoons unsalted butter
½ onion, finely chopped
3 cloves fresh garlic, minced
1-inch-piece fresh ginger, peeled and grated
2 teaspoons ground cumin
2 teaspoons paprika
1 teaspoon ground turmeric
1 big pinch of cayenne
1 teaspoon salt

¼ teaspoon freshly ground black pepper
2 tablespoons honey
28 ounces crushed tomatoes
½ cup Homemade Chicken Stock (page 79)
4 cups chopped cooked chicken
¼ cup half-and-half
Juice of 1 lemon
3 cups cooked rice

Directions

Melt the butter in a large pan over medium-high heat. Add onion, garlic, and ginger and sauté for 3 to 4 minutes until the onion is translucent and fragrant.

Add the cumin, paprika, turmeric, cayenne, salt, and pepper. Stir well, then add honey, tomatoes, and stock. Bring to a gentle boil, reduce heat to low, and simmer for 10 minutes.

Add chopped chicken and stir to combine.

Remove from heat and stir in half-and-half and lemon juice.

Serve over cooked rice with a fresh salad.

Chicken Caprese Sandwiches

Chicken Caprese Sandwiches feature all the best fresh summer flavors on a sandwich and make a great quick dinner or perfect summer lunch! Serve with a tossed salad and enjoy.

Serves: 4 (2 open-faced sandwiches per person)
Prep time: 15 minutes

Ingredients

4 ciabatta rolls
2 cups shredded cooked chicken
Juice from ½ a lemon
1 teaspoon extra-virgin olive oil
2 tablespoons chopped fresh parsley (or 2 teaspoons dry)
1 tablespoon chopped fresh oregano (or 1 teaspoon dry)
¼ teaspoon salt
Freshly ground pepper to taste
8 ounces fresh mozzarella, sliced into 8 pieces
4 tomatoes, sliced into 16 thin slices
Fresh basil leaves
Balsamic vinegar to taste

Directions

Cut four ciabatta rolls (or a loaf of sourdough bread if preferred) in half and lightly toast. If desired, even the top of the ciabatta rolls by slicing off the top so they lay flat. Save the tops for homemade bread crumbs or feed them to the chickens.

Mix chicken, lemon juice, olive oil, parsley, oregano, salt, and pepper and set aside.

To make the Chicken Caprese Sandwiches, spoon chicken over the toasted bread, and top each slice with one piece of mozzarella, two small tomato slices, and fresh basil leaves.

Drizzle with balsamic vinegar and add more salt and pepper to taste.

Chicken Bacon Ranch Flatbread Pizza

Another recipe perfect for a quick lunch or dinner, this flatbread pizza is fast and delicious. It features leftover chicken, homemade ranch dressing, and pre-bought flatbread. Get a quick meal on the table in under 20 minutes with this recipe. Serve with a fresh salad and enjoy!

Serves: 8
Prep time: 10 minutes
Cook time: 10 minutes

Ingredients
2 rectangular premade pizza crusts, 7½ ounces each
2 tablespoons extra-virgin olive oil
2 cloves garlic, minced
2 cups chopped cooked chicken
1 cup Homemade Ranch Dressing (page 94)
2 cups shredded cheddar cheese
½ cup chopped cooked bacon
Sliced green onions, optional

Directions
Preheat oven to 425°F.

Lightly brush top of both crusts with extra-virgin olive oil and half the crushed garlic. Put half the cooked chicken, ranch dressing, cheese, and bacon on each pizza crust.

Bake directly on the oven rack for 5 to 8 minutes for a crispier pizza, or on a pizza stone for 8 to 12 minutes.

Garnish with sliced green onions, if desired.

Let cool for 5 minutes, slice, and enjoy.

Chicken Potpie from Scratch

There's really nothing better than homemade chicken potpie from chicken and vegetables you raised yourself. This recipe makes a hearty and delicious meal. It also freezes well, too, so double it when you serve this for dinner and put an extra potpie in the freezer.

Serves: 6
Prep time: 25 minutes
Cook time: 45 to 60 minutes

Ingredients

Double piecrust (page 126)
1 cup Homemade Chicken Stock (page 79)
1 cup diced potatoes (1–2 medium potatoes)
1 cup chopped carrots (1–2 medium carrots)
1½ tablespoons unsalted butter

1½ tablespoons flour
1 cup chopped cooked chicken
½ cup frozen peas
¼ cup milk
Salt, to taste
Pepper, to taste
1 egg

Directions

Preheat oven to 350°F.

Prepare a 9-inch pie pan with the bottom piecrust. Poke a few holes in the crust with a fork so that it won't bubble.

Bring chicken stock to a rolling boil, add potatoes and carrots, and cook for 6 to 8 minutes over medium heat.

Prepare a roux by melting butter in a medium saucepan over medium heat. Whisk flour into the butter and add milk. Whisk until thickened and bubbly.

Pour the gravy into the potato and carrot mix. Add the chopped chicken and frozen peas. Stir until combined. Add salt and pepper to taste and pour into the prepared pie shell.

Make an egg wash by whisking egg with 1 tablespoon water. Add the top crust, make slits in the top, and brush with egg wash.

Bake 45 minutes or until brown and bubbly. Remove from oven. Let rest 5 to 10 minutes. Slice and enjoy!

Great-Grandma Shafer's Piecrust Recipe

I first started making this delicious piecrust when I wanted to re-create my grandma's strawberry shortcake. She always used piecrusts instead of shortcake, and everyone loved it. Her recipe for piecrusts remains my favorite to this day.

Yields: 1 double pie crust
Prep time: 25 minutes

Ingredients

1 teaspoon salt
2 cups all-purpose flour, unbleached
1 cup coconut oil shortening (or lard like my great-great-grandma used)
¼ cup cold water + 1 tablespoon if necessary

Directions

Mix salt with flour. Cut in shortening (or lard) with the edge of a spoon until all the flour is used and the mixture resembles crumbles.

Sprinkle water on top, and stir until the mass holds together.

Place on floured board and knead lightly just until the dough can be rolled. Work in as little extra flour as possible.

Note: It is easier to work a cold piecrust, so if you're having trouble rolling this out, cool it off in the refrigerator for 20 minutes before rolling.

Taco Seasoning Blend

This quick spice blend is so versatile and easy to make. It's used in the Chicken Fajita Sheet-Pan Dinner (page 113), Chicken Taco Zucchini Boats (page 117), and Chicken Burrito Bowls (page 111). I always make it in a time-saving big batch because we use this blend so much.

Yields: ⅔ cup
Prep time: 5 minutes

Ingredients
¼ cup chili powder
2 tablespoons ground cumin
1 tablespoon salt
1 tablespoon black pepper
2 teaspoons granulated garlic powder
2 teaspoons onion powder
2 teaspoons dried oregano
2 teaspoons paprika
¼–½ teaspoon crushed red pepper flakes

Directions
Mix all spices together in an airtight jar. Shake well to combine. For a less spicy mix, only use ¼ teaspoon crushed red pepper flakes.

Spicy Blackened Seasoning Blend

Great to use on chicken and fish, this spice blend is used in the Blackened Chicken Tacos (page 108). It's a little on the spicy side, so omit red pepper flakes if you can't take the heat.

Yields: ½ cup
Prep time: 5 minutes

Ingredients
2 tablespoons ground paprika
1 tablespoon cayenne
1 tablespoon onion powder
1 tablespoon garlic powder
2 teaspoons salt
1 teaspoon pepper
½ teaspoon crushed red pepper flakes
½ teaspoon oregano
¼ teaspoon thyme
¼ teaspoon basil

Directions
Mix all spices together in an airtight jar. Shake well to combine. For a less spicy mix, only use ½ teaspoon crushed red pepper flakes.

Rotisserie Spice Blend

This delicious mix of spices makes a lovely homemade rotisserie chicken to rival all store-bought rotisserie chickens! This recipe is used in Rotisserie Chicken (2 Ways) and can also be used to spice up chicken noodle soup.

Yields: ¾ cup
Prep time: 5 minutes

Ingredients
2 tablespoons chili powder
2 tablespoons salt
2 tablespoons black pepper
2 tablespoons garlic powder
2 tablespoons onion powder
2 tablespoons dried thyme
2 tablespoons dried paprika

Directions
Mix all spices together in an airtight jar. Shake well to combine.

Greek Seasoning Spice Blend

If you like fresh Greek flavors, this spice blend is for you! Like the other spice blends, it's very easy to make and keep on hand for chicken to use in salads, breakfast muffins, and grilled chicken, too.

Yields: 1 cup
Prep time: 5 minutes

Ingredients
3 tablespoons oregano
2 tablespoons basil
2 tablespoons onion powder
2 tablespoons garlic powder
2 tablespoons salt
1 tablespoon dill
1 tablespoon parsley
1 tablespoon rosemary
1 tablespoon marjoram
1 tablespoon ground black pepper

Directions
Mix all spices together in an airtight jar. Shake well to combine.

Glossary of Terms

Here are definitions to a few terms you will read about repeatedly throughout this book. You might already know what the mean, but if you come across a term you aren't familiar with, hopefully this section will help.

Broiler, Fryer, Roaster: These terms are all used interchangeably to mean a young chicken that is being raised only for meat.

Brooder: The brooder is a warm, draft-free, and dry container that will first house your chicks.

Broody hen: A broody hen is a hen trying to hatch eggs or raise chicks.

Cockerel: The opposite of a pullet, a cockerel is a young male chicken.

Crop: Also called a *craw*, the crop is part of the chicken's digestive system. It's where the food is held initially before moving further down the digestive path.

Eviscerate: To *eviscerate* or *the process of evisceration* means to take out the entrails. During the evisceration process of chicken butchering, the intestines and internal organs are removed.

Gizzard: Another part of the chicken's digestive system. The gizzard is where the chicken's food is ground up by using small stones or grit.

Heritage breeds: Unlike chickens bred for confinement, heritage breeds are types of chickens that have been around a long time. They tend to be more versatile and hardier in homestead environments.

Killing cone: A cone used for humane chicken butchering. A killing cone holds a chicken upside down and hugs their bodies snugly, allowing easy access to the chicken's neck.

Pullet: Simply put, a pullet is a young female chicken.

Spatchcock: Spatchcock is a cutting technique that involves removing a chicken's backbone and flattening the whole chicken. It helps speed up the cooking process and makes the cooking time more uniform.

Straight run: The most economical way to buy chickens for meat, a *straight run* simply means a mix of pullets and cockerels. It's the luck of the draw when you order straight-run chicks.

Vent: The meat chicken's vent is where the waste exits the chicken's body. It's also where eggs exit.

About the Author

Michelle Marine, recovering city girl, lives on five acres in eastern Iowa with a houseful of kids and a homestead full of animals. She started the popular lifestyle blog *Simplify, Live, Love* in 2010 to encourage families to lead more eco-conscious lives and to enjoy each other more by focusing on what matters: local food, family fun, and travel. Michelle has a snarky sense of humor, an undergraduate degree in German, and a master's degree in English Linguistics. First as a child, and later as a young adult, Michelle spent twelve total years living abroad in Europe and Asia, which fueled her passion for green living. Her interest in homesteading comes from her Grandma Ruth Shafer, who grew her own food, raised her own chickens, and home-birthed six children long before it was the cool thing to do. Michelle loves Instagram, Great Pyrenees dogs, backyard fowl, solar power, passive houses, and aspires someday to be Barbara Kingsolver and eat only locally produced food. Her mixed flock of backyard birds is Instagram famous and can be found at @simplifylivelove.

Acknowledgments

While I get author credit for writing this book, so many people helped me behind the scenes to make it a reality. It takes a huge team of people to make a project like this possible, and I owe so many people a heartfelt thank-you!

Carissa, many thanks for recommending me for this project in the first place! Without you, it would not have been possible at all.

A special thank-you to my children—Anna, Ben, Cora, and Sara—for amusing themselves this summer so I could focus on the book. Thank you also for helping feed, water, and take great care of our meat chickens and other homestead animals.

Many thanks to my parents for feeding, chauffeuring, and being there when I need them. Also, thank you both for raising me to believe in myself and my abilities!

Robin, thank you for your friendship, cooking expertise, and taste-testing abilities. Thanks also for teaching me how to cut up a whole chicken. You are a great friend.

Nicky, thank you also for your friendship, for contributing your beer can chicken recipe, and for me telling me to get to work when I was procrastinating horribly. I couldn't have done this without you.

Sarah, thank you for butchering chickens with me. I appreciate your willingness to jump in there and help me get the job done! You are impressive, my friend!

To Jake and Heather, thanks for opening your home this summer while we were on vacation so I could finalize the draft at the last minute.

Janet, a huge thank-you for your encouragement and words of wisdom. As a seasoned homesteading author, you helped so much! I will be forever grateful.

Alicia and Hoover's Hatchery, a big thank-you for always providing amazing chicks! Without them, this book would have been a lot more difficult. To my editor at Skyhorse, Nicole Frail, thank you for holding my hand and helping me throughout the entire process! It wasn't nearly as hard as I thought it might be.

To my husband Dan, thank you for being there every step of the way: helping me clean up the kitchen, not complaining about chicken for dinner night after night, helping me butcher chickens, and most importantly, proofing my work to keep me from embarrassing myself. You made every step better and deserve so much credit.

But most of all, thank you to all of you! Without the encouragement of so many friends, social media followers, and blog readers, this book would not have been possible. I hope it gives you the confidence to try something new and to take ownership of a long-forgotten survival skill: how to raise your own food and provide for your family.

Index

apple cider, 49
Blackened Chicken Tacos
with Slaw & Lime Sauce,
109
Homemade Chicken Stock,
79–81
balsamic
Roasted Tomato &
Blueberry Salad with
Chicken, 95

W
walnuts, 95
waste, of feed, 46
water containers, 32–33
weighing, 53
withholding of feed, before
butchering, 57

Y
yogurt
Blackened Chicken Tacos
with Slaw & Lime Crema,
109
Greek Grilled Spatchcock
Chicken, 105

Z
zoning laws, 6–7
zucchini
Chicken Taco Zucchini
Boats, 117

Conversion Charts

METRIC AND IMPERIAL CONVERSIONS

(These conversions are rounded for convenience)

Ingredient	Cups/Tablespoons/ Teaspoons	Ounces	Grams/Milliliters
Butter	1 cup/ 16 tablespoons/ 2 sticks	8 ounces	230 grams
Cream cheese	1 tablespoon	0.5 ounce	14.5 grams
Cornstarch	1 tablespoon	0.3 ounce	8 grams
Flour, all-purpose (gluten-free)	1 cup	5.2 ounces	148 grams
Fruit, dried	1 cup	4 ounces	120 grams
Fruits or veggies, chopped	1 cup	5 to 7 ounces	145 to 200 grams
Fruits or veggies, pureed	1 cup	8.5 ounces	245 grams
Honey, maple syrup, or corn syrup	1 tablespoon	0.75 ounce	20 grams
Liquids: cream, milk, water, or juice	1 cup	8 fluid ounces	240 milliliters
Oats	1 cup	5.5 ounces	150 grams
Salt	1 teaspoon	0.2 ounce	6 grams
Spices: cinnamon, cloves, ginger, or nutmeg (ground)	1 teaspoon	0.2 ounce	5 milliliters
Sugar, brown, firmly packed	1 cup	7 ounces	200 grams
Sugar, white	1 cup/1 tablespoon	7 ounces/ 0.5 ounce	200 grams/ 12.5 grams
Vanilla extract	1 teaspoon	0.2 ounce	4 grams

OVEN TEMPERATURES

Fahrenheit	Celsius	Gas Mark
225°	110°	¼
250°	120°	½
275°	140°	1
300°	150°	2
325°	160°	3
350°	180°	4
375°	190°	5
400°	200°	6
425°	220°	7
450°	230°	8

Notes

Notes

Notes

Notes

Notes

Notes

Notes

Notes

Notes

Notes

Notes

Notes

Notes

Notes

Notes

Notes